Atlas of Design

Vol. 7 »

editors
Aileen Clarke
Daniel Coe
Erin Greb
Aaron Koelker

published by
nacis
the North American
Cartographic Information Society

Atlas of Design, Volume Seven

First Printing, October 2024

Editors: Aileen Clarke, Daniel Coe, Erin Greb, Aaron Koelker

Published by the North American Cartographic Information Society, *nacis.org*

The type in this book has been set in Avenir, by Adrian Frutiger.

Printed by The John Roberts Company, Coon Rapids, MN, and bound by Midwest Editions, Minneapolis, MN.

nacis

The *Atlas of Design* is a biennial NACIS publication. The North American Cartographic Information Society improves communication, coordination, and cooperation among the producers, disseminators, curators, and users of cartographic information. The organization is founded on education and information sharing, as well as promoting cartographers' works through publications such as the *Atlas*.

NACIS hosts an annual conference for professionals and enthusiasts from government, commercial, and not-for-profit groups along with academics, scholars, map/GIS librarians, artists, technologists, and students interested in the world of maps and geographic information. To learn more about NACIS visit nacis.org.

A Note from the Editors

This seventh volume of the *Atlas of Design* contains thirty-three unique maps (thirty-seven, if we want to get technical) and continues the series mission of promoting the understanding and appreciation of everything that goes into a well-crafted map. And while we're inclined to think that this book *is* quite beautiful, it is not just a collection of pretty pictures. Whether you read maps for pleasure, use them to accomplish tasks or stay informed, or are a cartographer yourself—we hope that the maps and commentaries in this book help to boost your understanding of the decisions behind intriguing cartography. Because by understanding these choices, we can not only better communicate with maps, but with all things that require the thoughtful intentions of design. From art and technology to policy and infrastructure—unraveling the logic behind these decisions ultimately helps us to communicate better with each other.

As always, we'd like to thank the hundreds of contributors who submitted their work for consideration in this volume. Without you, the *Atlas of Design* could not exist. While we can only feature so many designs in each book, we recognize and appreciate the extraordinary amount of time and effort put into each and every map that is sent our way, and we are heartened to see the medium so alive and well.

Helping us sort through all those maps was our panel of knowledgeable judges. Thanks go to Aly Ollivierre, Anton Thomas, Bill Limpisathian, Daniel Huffman, Delphine Papin, Jack Jen Gieseking, Jennifer Mapes, Jim Meacham, Jug Cerovic, Julie Witmer, Molly O'Halloran, and Tim Wallace. Without you, we'd probably still be on a video call debating over what to include.

We'd like to acknowledge the North American Cartographic Information Society, both board and body, for continuing to support the *Atlas*. While a lot of things have changed since the first volume was published, their support has remained constant and makes this series possible. We'd also like to acknowledge the previous editors of the series, upon whose work our efforts are built, as well as Nat Case, our series managing editor, who made our lives much easier by handling the many less fun aspects of maintaining a book series.

We also would like to thank Lys Weiss of Post Hoc Academic Publishing Services for helping to keep yet another volume sounding clear and consistent, as well as our printing partners for this volume—The John Roberts Company and Midwest Editions—for helping to bring the book to physical fruition.

And of course we thank you, our curious reader. Whether you purchased this book yourself, received it as a gift, or borrowed it from a library or teacher—we appreciate your interest in cartographic design and hope that, in these maps, you'll find something new to admire in the world.

Introduction by Nat Case

The maps in this volume are amazing. I can say this confidently even though I haven't seen them as I write this essay. All the volumes of the *Atlas of Design* are full of stunning specimens of cartography—the editors' challenge is not finding terrific maps, but deciding which ones to include. I speak from experience.

What I'm thinking about now, though, is the design of other maps—the maps we cartographers make day in and day out. This week I'm working on a basic parks and trails map for the local park board, maps for a US history atlas (I just finished "Irrigation in the 1930s"), and a map showing 1849 land claims in St. Paul, Minnesota. I'm happy with them, but they are not the kinds of maps you'd find in the *Atlas*, because their design is, well, *ordinary*.

Mapmaking, like most making, is largely about solving problems. You have *this* area of coverage and *that* space to fill. You will be using *this* display technology, and need to work within *that* style specification. You have *this* collection of information, and need to sort it into *that* many categories. Given all that, your problem is how to make your map communicate what it needs to, clearly and appealingly. That's map design. It's what cartographers do every day. It's a job.

I was a studio art major in college, back when mammoths roamed the earth. In the short time I spent on the edges of the art world, I noticed that a lot of emphasis was placed on things like "uniqueness" and "finding your distinctive voice." The unique expressive self is central to how we frame the fine arts. The art marketplace is dominated by the creation of unique (and therefore precious) objects. But what makes something unique, valuable, or distinctive? What makes it *original*?

Maybe originality is overrated.

Don't get me wrong. I love seeing someone reveal a new, never-before-seen perspective on the world. I love when I can see someone's own skilled, personal hand in their map—I'm a sucker for hand-drawn, hand-carved, or hand-printed maps. But that's not what makes a map "good." Not on its own, anyway.

Terrific ordinary maps exist because cartography is a *tradition*. I know the word comes with a lot of baggage—people wanting to maintain social stasis use it a lot—but I want to push back on that baggage. Tradition is a *context*, a set of conventions and standards and assumptions that support an activity across generations. The continuity between maps is made of continuously shared (and sometimes shamelessly borrowed) techniques, language, and style. It's the commonality that tells us "that's a map!"—an experience that Mark Denil has described as "mappicity."

We make maps that *look like maps*. And everyday cartography—the look and feel of *all* maps in our tradition—is key to that continuity.

This is something I love about the field, and part of what I love about the *Atlas of Design*. I guarantee there are at least a couple of maps in this volume that are direct tributes to older maps and map styles. And there are a lot more that I'll look at and say, "I know where they got the idea!"—for a font, color palette, or shading technique.

My calling cartography a "tradition" doesn't come out of nowhere. I have been a Morris Dancer—Morris is a type of traditional English folk dance—for about as long as I've been making maps. My group dances a particular style based on what was danced in a particular English village as collected by folklorists about 125 years ago. But we have made that style distinctly our own, and over the life of my team (thirty-six years) that style has evolved to look very little like what was danced a century ago. We have choreographed new dances that make up most of our current repertoire, many of them set to new tunes. So, is what we do "tradition"? Heck, yes! When we make new dances, they *look* like Morris dances. We work within basic forms and patterns, coming up with new ways of arranging them, sometimes adding a piece that was simply not there before. It is, as they say, a *living* tradition.

What makes a good Morris dance, what makes a good map—what makes a good anything—*does* include originality, but an originality that might not be rule-breaking or "never seen before." It's an originality that knows the mental and physical tools of the craft, and works within the broad shape of the tradition to make something new.

What is new and fresh about the maps in this book? Again, even without seeing them, I can tell you that some of the freshness comes from the appearance of someone's "voice"—someone clearly, personally speaking through the map. Some of it comes from addressing an aspect of the world we didn't know before. Some of it comes from new techniques—our tools are constantly evolving, as are the ways people find to combine and leverage them. And some of it is simply this: a mapmaker who uses tools we've seen before, who depicts something we've seen before, in a voice we've heard before, and gives us a gorgeous window into the world. We blink . . . and our familiar world has been portrayed anew. Someone recites a very old sonnet, and our ears are opened in a way they were not before.

This is one of the magical things about how people depict the world. Even if the point of the depiction is one you've seen or heard a thousand times, if you tell it right, a love song or a portrait or an action movie can still take your breath away. And even the maps that look like maps you've seen before, when they work right, can do that too. We hear a familiar tune, we see that same old story . . . and the map is *still* fresh and glorious, and we just want to dive in and explore it.

Traditions aren't pressed flowers. They aren't stuffed animals in a museum. Traditions *live*. And like all living things, they grow and change, beget progeny, and eventually fade and die. Cartography is very much alive. It's learning and growing, as all of us cartographers are learning and growing, and *that* is what this book series is all about. Not originality or novelty alone, but about the living, thriving craft of making maps.

Arctic Ocean
KALAALLIT NUNAAT
GREENLAND
Hudson Bay
Beaufort Sea
Barents Sea
NORTH AMERICA
North Pacific Ocean
North Atlantic Ocean
CENTRAL AMERICA
SOUTH AMERICA
South Atlantic Ocean
AFRICA
SAHARA
ARABIAN PENINSULA
Indian Ocean
Wild World
Southern Ocean
ANTARCTIC PENINSULA
Weddell Sea
Queen Maud Land
WEST ANTARCTICA
ANTARCTICA
EAST ANTARCTICA

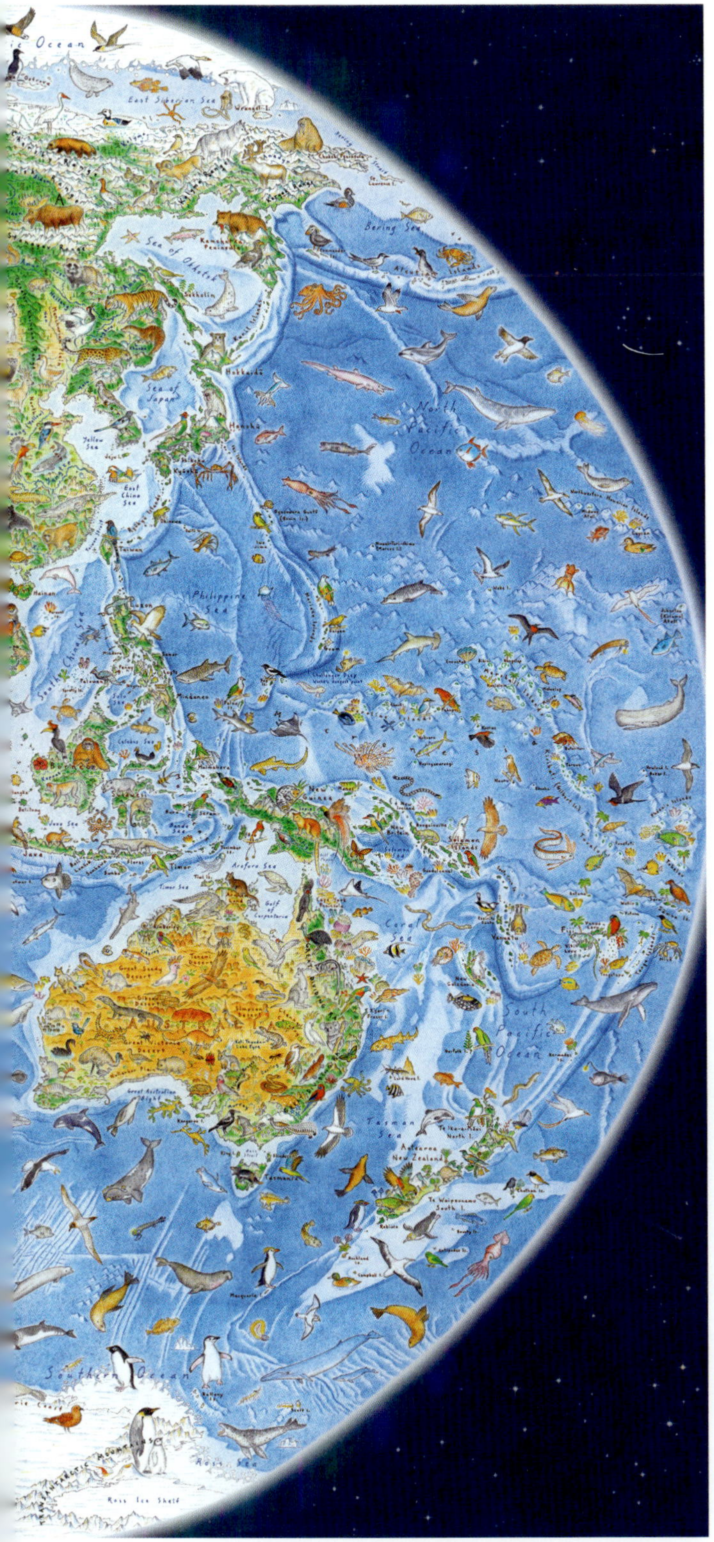

« Wild World

Anton Thomas

"Wild World" is a celebration of Earth's geography and wildlife. It has no countries or cities, but hosts 1,642 animals. It took me three years (full-time) to complete the map, which is entirely drawn with colored pencils and pen.

My love of drawing maps began in childhood, amid the mighty geography of Aotearoa's South Island. A parallel passion for wildlife had me featuring animals on my maps from the start. Fast-forward to 2020, when with a yearning to escape the confines of a global pandemic, I decided to commence a lifelong dream: a world map of nature.

I chose the Natural Earth projection, printing a base map and stapling it to the back of the drawing paper. This allowed me to trace all the basic geography. "Wild World" is not just about wildlife—it's a very detailed physical world map. A high level of accuracy was vital.

With coastlines traced and pencils sharp, the magic could begin. I started in Alaska, and a grizzly bear with a mouthful of salmon was the very first animal I drew. I soon settled on three criteria that shaped the philosophy of the map. All 1,642 animals on "Wild World" are (1) wild, (2) native, and (3) extant (not extinct).

Wild and native animals invoke natural biomes, while extant animals set the map in the present day. This is crucial: "Wild World" is a place that still exists, a world that can still be cherished and protected. I wanted this map to inspire hope, to show how much of the wild world is still left to preserve.

"Wild World" required a vast breadth of research and thousands of drawing hours. It was extremely challenging work, but I loved it. Exploring the world like this was magical. Animals I never dreamed existed were introduced to me. Landscapes of profound beauty were revealed to me.

Ultimately, I wanted to portray the Earth in a less anthropocentric way. To take the spirit of a good physical map, combine it with wildlife, and craft a new perspective. A world map of nature, rather than nations. I hope "Wild World" reminds people how amazing this planet is and that it's worth fighting for.

Yukon
Alaska Range
Mackenzie Mts.
Great Slave Lake
Barren Grounds
Thelon
Southampton I.
Hudson Strait
Hudson Bay
Ungava Peninsula
Kenai Pen.
Kodiak I.
Gulf of Alaska
Liard
Peace
Athabasca
Churchill
Coast Mountains
Rocky Mountains
Haida Gwaii
Fraser
Saskatchewan
Nelson
Lake Winnipeg
Vancouver Island
Columbia
Missouri
Great Plains
L. Superior
Laurentian Mountains
Cape Breton
Yellowstone
Cascade Range
Snake
L. Michigan
L. Huron
St. Lawrence
Sable I.
L. Ontario
Gulf of Maine
Great Basin
Platte
Ohio
Long I.
Sierra Nevada
Colorado
Appalachian Mountains
North America
Ozark Plateau
Arkansas
Channel Is.
Gila
Sonoran Desert
Mississippi
Red
Alabama
Bermuda
Guadalupe I.
Baja California
Chihuahuan Desert
Rio Grande
Florida
Bahamas
Gulf of México
Andros I.
Sierra Madre Occidental
Sierra Madre Oriental
Revillagigedo Is.
Cuba
Bay of Campeche
Yucatán Pen.
Cayman Is.
Gulf of Honduras
Jamaica
Hispaniola
Puerto Rico
Caribbean Sea
Guadeloupe
Lesser Antilles
Barbados

Labrador Sea
Iceland
Faroe Is.
Shetland Is.
Rockall
North Sea
Great Britain
Ireland
Baltic Sea
Baltic Plains
Daugava
Lake Onega
Lake Ladoga
Moscow
Vistula
Oder
Elbe
Rhine
EUROPE
Dnieper
Seine
Loire
Danube
Carpathian Mountains
Dniester
North Atlantic Ocean
Bay of Biscay
ALPS
Po
Rhone
Dinaric Alps
Danube
Crimea
Black Sea
Duero
Pyrenees
Iberian Peninsula
Tagus
Corsica
Sardinia
Balearic Is.
Italian Peninsula
Adriatic Sea
Balkan Peninsula
Anatolia
Azores
Ionian Sea
Sicily
Malta
Mediterranean Sea
Crete
Cyprus
ATLAS MOUNTAINS
Saharan Atlas
Madeira
Anti Atlas
Grand Erg Occidental
Grand Erg Oriental
Canary Islands
Great Sand Sea
Sinai
Libyan Desert
Nile
Erg Iguidi
Erg Chech
Idhan Murzuq
Ahaggar Mtns.
SAHARA
El Djouf
Ténéré
Tibesti Mtns.
Air Mtns.
Cape Verde Islands
Sénégal
Gambia
Niger
Benue
SAHEL
Lake Chad
Nile

Wayfinding Map for Calgary's +15 Skywalk »

David Kopulos, James Brown, Clare Seldon, Helen McKenzie, Emily Whiteside, Sarah Yuksel, Alex Syniukov, Nasrin Veiga, and Juan Rioseco, Steer

Ask a Calgarian about the downtown "+15 Skywalk" network—so named because it stand 15 feet above street level—and they'll tell you: it's handy during the long, cold winters, but not an easy place to navigate. This reputation has discouraged people from using +15, impacting the retailers along its 16 km of walkways and preventing its 86 bridges from serving as the "elevated city" it was intended to be.

Our challenge, therefore, was not just to design a map to improve wayfinding, but also to reshape +15's reputation and residents' mental map of the Skywalk.

Rather than designing a detailed map that depicted the network's every twist and turn, we opted to portray +15 in the approachable style of a transit diagram. The map simplifies +15 to its essence and answers users' most important questions: Where am I, and what destinations can I reach in each direction? Supported by new on-site signage that will provide clarity in complex areas, the map establishes a new image of +15 as an inviting place worth exploring.

To execute this schematic diagram, we developed a graphic style that reflects the city's "modern cowboy" culture. Whimsical touches, such as a palm tree marking a public greenhouse and the city's iconic white Stetson hat indicating your location, give the map a unique sense of place and a welcome sense of humor.

The map's cool, bold palette is an homage to the original +15 brand, developed in the 1980s by legendary designer Lance Wyman. High-contrast colors aid low-vision viewers, and the linework suggests an x-ray image, as if revealing the hidden layers of the city.

To assist people in planning stair-free journeys, the map shows the locations of stairways and accessible alternate routes. More than 500 residents provided comments on the draft map, which helped us to understand how people saw +15 and what map features they valued most.

This map, now installed at 70 locations and regularly updated online to reflect temporary bridge closures, is at the core of a new wayfinding program for +15. We hope that it will become an iconic design for the next generation of Calgarians and an indispensable tool for navigating the city.

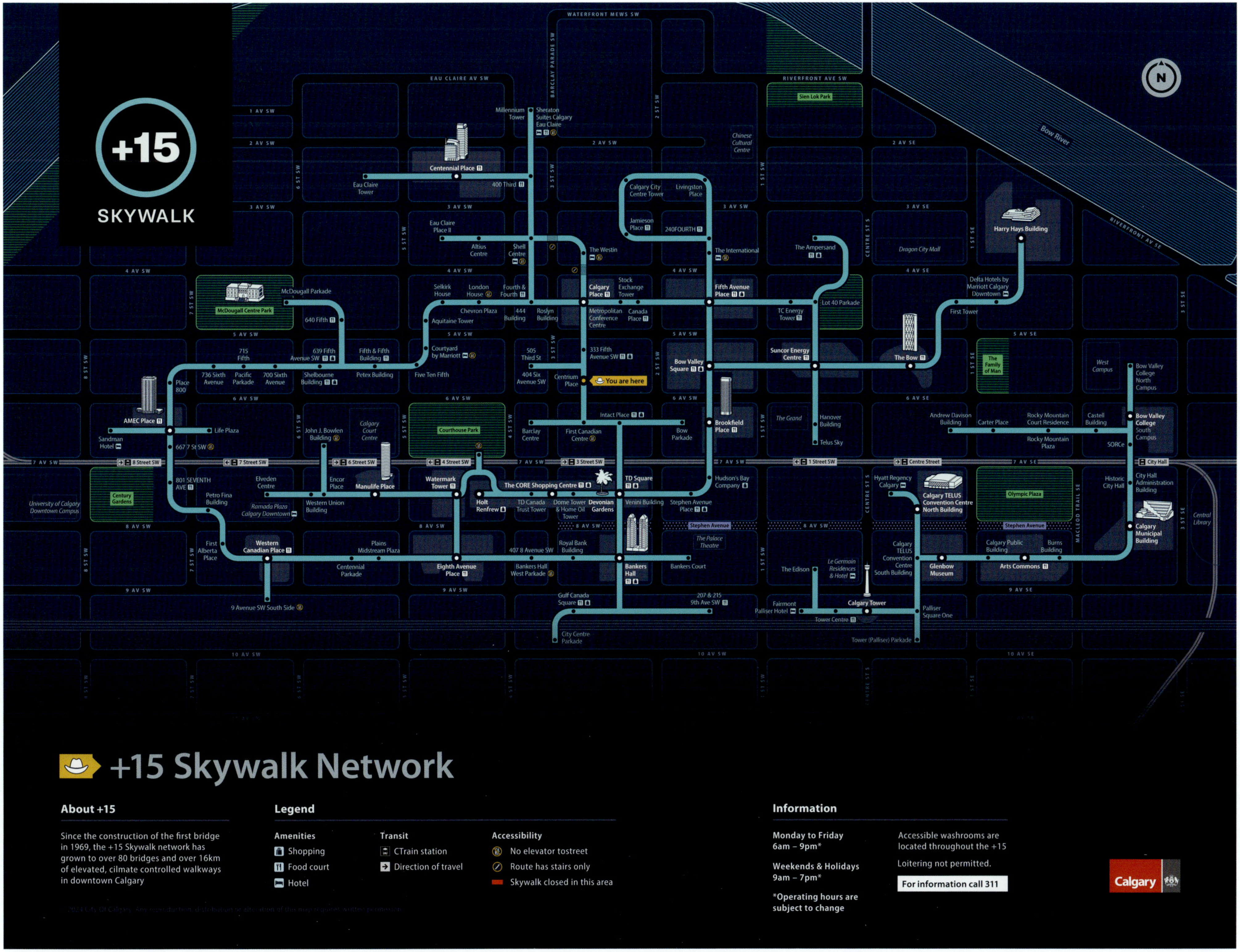

+15 Skywalk Network

About +15

Since the construction of the first bridge in 1969, the +15 Skywalk network has grown to over 80 bridges and over 16km of elevated, cilmate controlled walkways in downtown Calgary

Legend

Amenities

- Shopping
- Food court
- Hotel

Transit

- CTrain station
- Direction of travel

Accessibility

- No elevator tostreet
- Route has stairs only
- Skywalk closed in this area

Information

Monday to Friday
6am – 9pm*

Weekends & Holidays
9am – 7pm*

***Operating hours are subject to change**

Accessible washrooms are located throughout the +15

Loitering not permitted.

For information call 311

Calgary

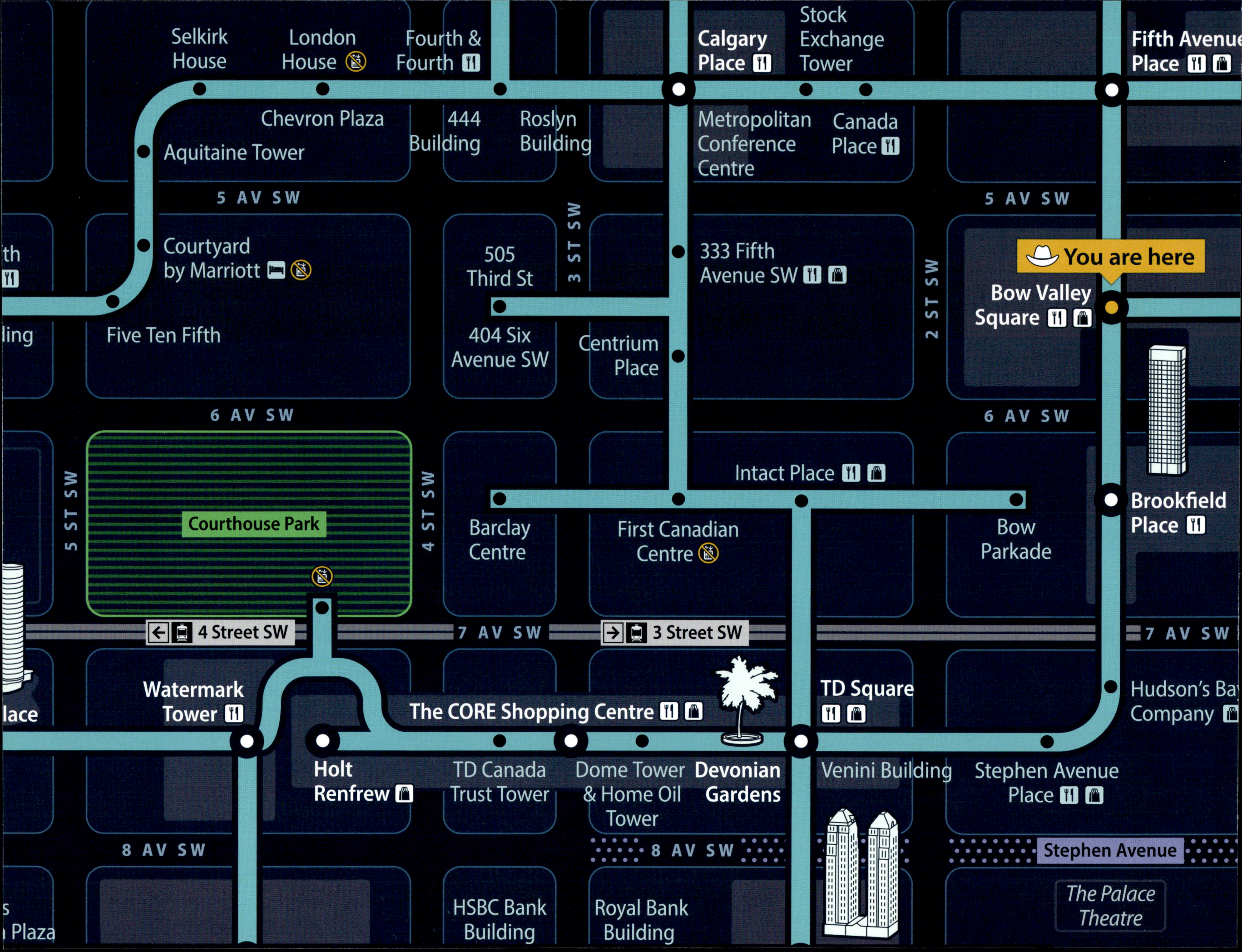
Selkirk House
London House
Fourth & Fourth
Calgary Place
Stock Exchange Tower
Fifth Avenue Place
Chevron Plaza
444 Building
Roslyn Building
Metropolitan Conference Centre
Canada Place
Aquitaine Tower
5 AV SW
Courtyard by Marriott
505 Third St
3 ST SW
333 Fifth Avenue SW
2 ST SW
You are here
Bow Valley Square
Five Ten Fifth
404 Six Avenue SW
Centrium Place
6 AV SW
Intact Place
Brookfield Place
5 ST SW
Courthouse Park
4 ST SW
Barclay Centre
First Canadian Centre
Bow Parkade
4 Street SW
7 AV SW
3 Street SW
Watermark Tower
The CORE Shopping Centre
TD Square
Hudson's Bay Company
Holt Renfrew
TD Canada Trust Tower
Dome Tower & Home Oil Tower
Devonian Gardens
Venini Building
Stephen Avenue Place
8 AV SW
Stephen Avenue
The Palace Theatre
HSBC Bank Building
Royal Bank Building

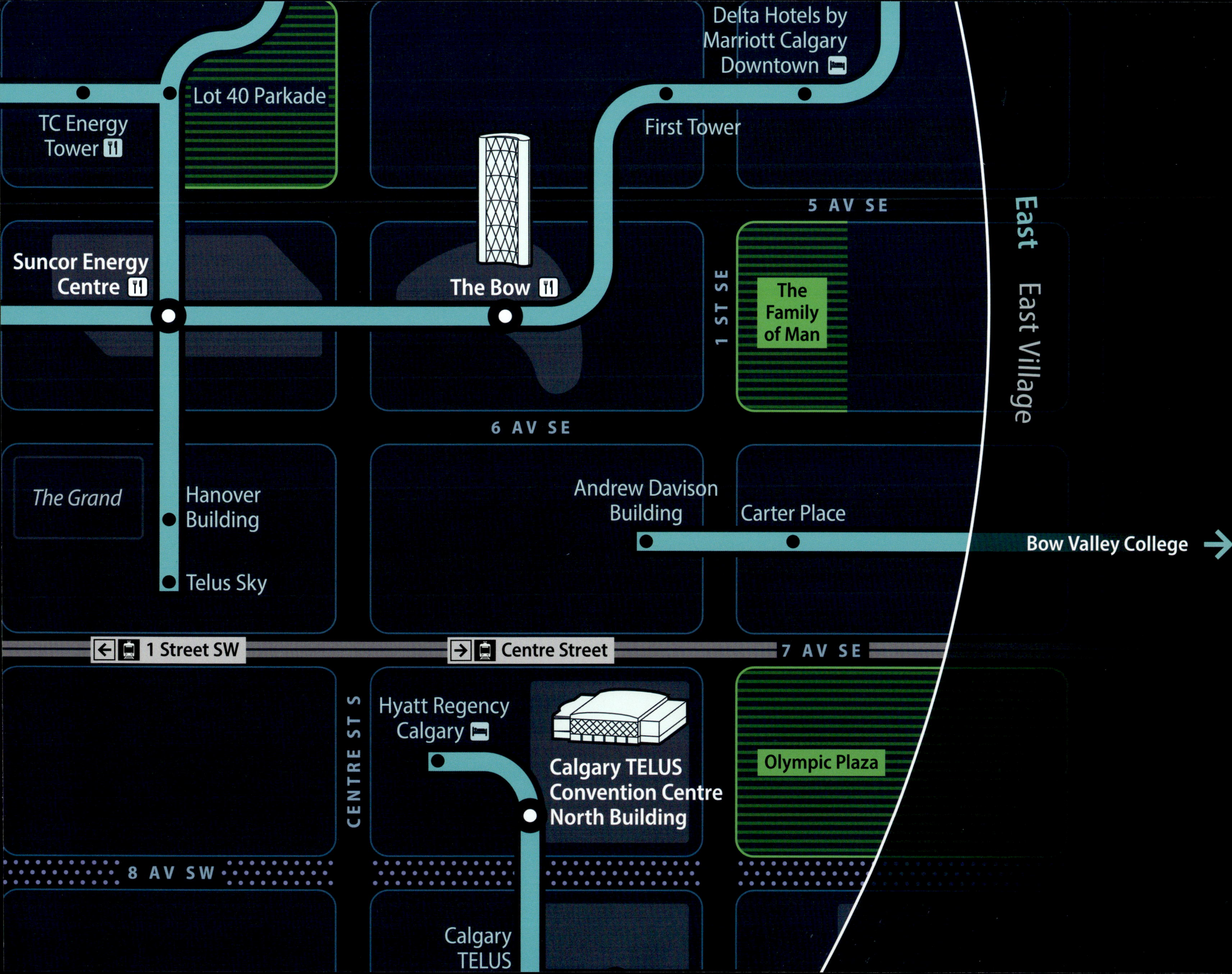

Delta Hotels by Marriott Calgary Downtown
Lot 40 Parkade
TC Energy Tower
First Tower
5 AV SE
East
East Village
Suncor Energy Centre
The Bow
1 ST SE
The Family of Man
6 AV SE
The Grand
Hanover Building
Andrew Davison Building
Carter Place
Bow Valley College
Telus Sky
1 Street SW
Centre Street
7 AV SE
Hyatt Regency Calgary
CENTRE ST S
Calgary TELUS Convention Centre North Building
Olympic Plaza
8 AV SW
Calgary TELUS

5x5-mile sections of famous U.S. National Parks, depicted above sea level & without vertical exaggeration
DENALI
MOUNT RAINIER
OLYMPIC
ZION
CRATER LAKE
YOSEMITE
CHANNEL ISLANDS
GRAND CANYON
HAWAI'I VOLCANOES
DEATH VALLEY

BADLANDS
ACADIA
YELLOWSTONE
ROCKY MOUNTAIN
BLACK CANYON of the GUNNISON
GREAT SMOKY MOUNTAINS
GRAND TETON
GREAT SAND DUNES
MESA VERDE
BISCAYNE
Created by Wendy Shijia Wang and RJ Andrews in 2024 using USGS and NLCD data.

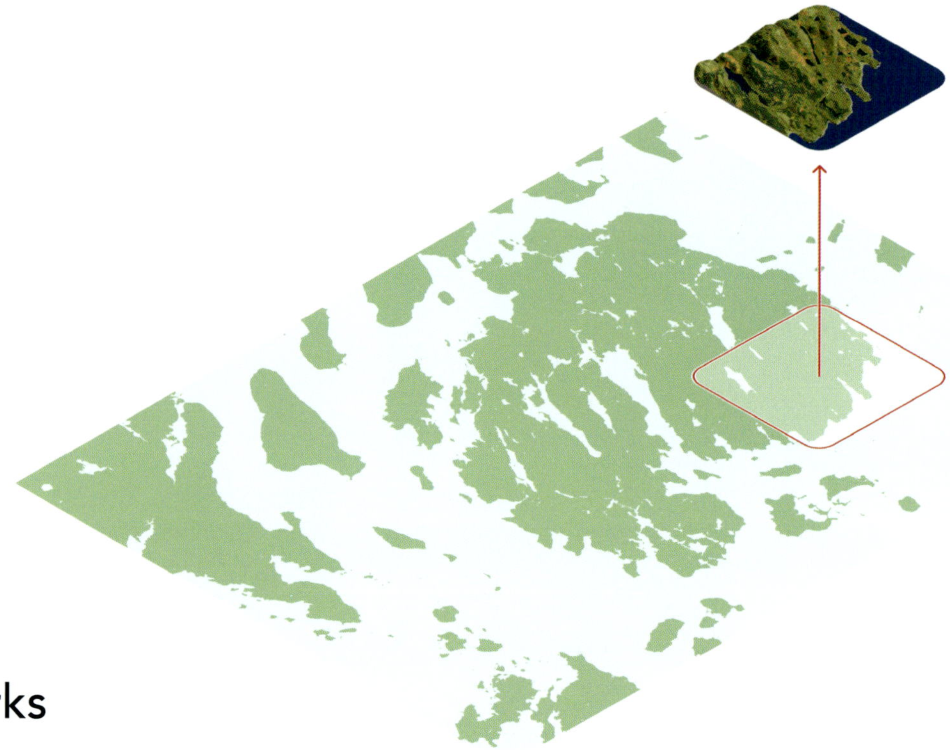

« A Delicious View of U.S. National Parks

Wendy Shijia Wang and RJ Andrews

Standing in Yosemite Valley, you might sense that it is taller and wider than Zion Canyon. But by how much? And are the two at the same elevation?

In 1972, Charles Eames described design as "an expression of purpose." The purpose of this composition is to enable natural comparisons between U.S. National Parks. Equal-area sections present features, elevations without exaggeration, and landform colors. Seeing any park in the context of the others further distinguishes what makes each one special. The Grand Canyon is somehow even more grand.

This project began as a free exploration of cartographic possibilities with emerging modeling tools. Wendy in Shanghai had already created many three-dimensional map experiments. RJ in San Francisco brought a desire to see the National Parks in a new way: What if each had a cookie-cutter section removed and placed on a chessboard for easy comparison?

Out of a total of sixty-three National Parks, we selected these twenty based on popular familiarity and topographic diversity. The design element that took the longest to develop was the style of the elevation models. Our goal was to convey the emotional appeal of each park—its vibe. We wanted something closer to how a midcentury poster would portray the park, and less how it photographs from an airplane. Indeed, we found that aerial imagery wasn't colorful enough (and removing its shadows was too troublesome). Instead, we relied on National Land Cover Database areas augmented with custom three-dimensional textures (for instance, tiny bumps for forests) and park-specific colors. Thus, the same "barren" data layer appears differently depending on which park you are in: medium gray in Rocky Mountain and orangey tan in Grand Canyon.

Our original intent and vision, however, was only part of our motivation over the many months of creation. Our maps gave us permission to learn from and relish each other's passionate design-thinking, shared inspiration, and technical know-how. David Rumsey says that "maps make my brain happy." To us, working through this effort together, while 6,000 miles apart, has made us both very happy. This project afforded us an opportunity to delight in joyful cartography as we spurred each other's enthusiasm for seeing the world in delicious ways.

"iDiscover Ding Ding": Hong Kong's Heritage Tram Routes »

Tania Willis, i-Discover and HK Tramways

I was commissioned by i-Discover and HK Tramways to illustrate a guide map of the heritage-rich neighborhoods lining Hong Kong's tram routes ("Ding Ding" is the double warning bell when the tram departs and also an endearing nickname given to it by Hong Kongers). I lived in Hong Kong for twenty years and had a bizarre love of these trams, so this was my dream project.

In my illustrations, I use imagery, storytelling, and aesthetics to convey a sense of place. Pictorial icons enhance the wayfinding and encourage viewers to explore the area. I like my icons to be recognizable but playful, and also to work as standalone images, as they are often used in other contexts (for example, the tram livery). I want to find the right balance: a beautiful map that has functional clarity, while retaining a strong sense of place. Aesthetically, color is a large part of the Hong Kong story, and I created a palette that reflects the city's characteristic vintage feel, deriving the core colors from the well-known green enamel that the trams have always used and the coral red often found on Chinese lacquerware.

As the map was commissioned for a concertina pamphlet, it had to work within a narrow space, which is not quite how the Hong Kong shoreline runs. So my challenge was to balance navigational demands with a kind of creative recomposing. The use of numbers allowed me to edit the icon content so that it didn't overwhelm the reader.

The later transposition of the map onto the tram livery was exciting. This was effectively an environmental advertising livery brief, meant to encourage people to download the associated map app when they saw the info on the tram. Cartographic accuracy was deprioritized, as the most important aspect was that it should be eye-catching. The most demanding aspect was the layout of the icons, ensuring they fitted in the positions where the bodywork allowed.

For me, the fundamentals of design involve solving problems with clear but expressive visual communication. It is a three-way conversation between the author of the problem, the designer (the problem-solver), and the end user.

ODS
iDISCOVER 叮叮
DING DING 叮叮
EXPLORE NEIGHB
ALONG THE TRAM
HKCEC
香港會議展覽中心
WAN CHAI
灣仔
WAN CHAI
灣仔站
Egg Tarts
檀島蛋撻
Star Street
星街
Pawn Shop
振安大押
安振
押
Wan Chai Market
舊灣仔街市
Herbal Tea
楊春雷涼茶
Milk Tea
Snake Wine
蛇王槳
Tai Temple
北帝廟
Noon Day
怡和午炮
CAUSEWAY BAY
銅鑼灣
HENNESSY RD
QUEEN'S RD EAST
CONNAUGHT RD CENTRAL
DES VOEUX RD CENTRAL
CENTRAL
中環站
ADMIRALTY
Bank of China
中銀大廈
Victoria Barracks
CENTRAL
ADMIRALTY
Fringe Club
藝穗會
Duddell St
Gas Lamps

172
iDiscover
#getlostwithoutgettinglost
i-discoverasia.com
APP×MAP

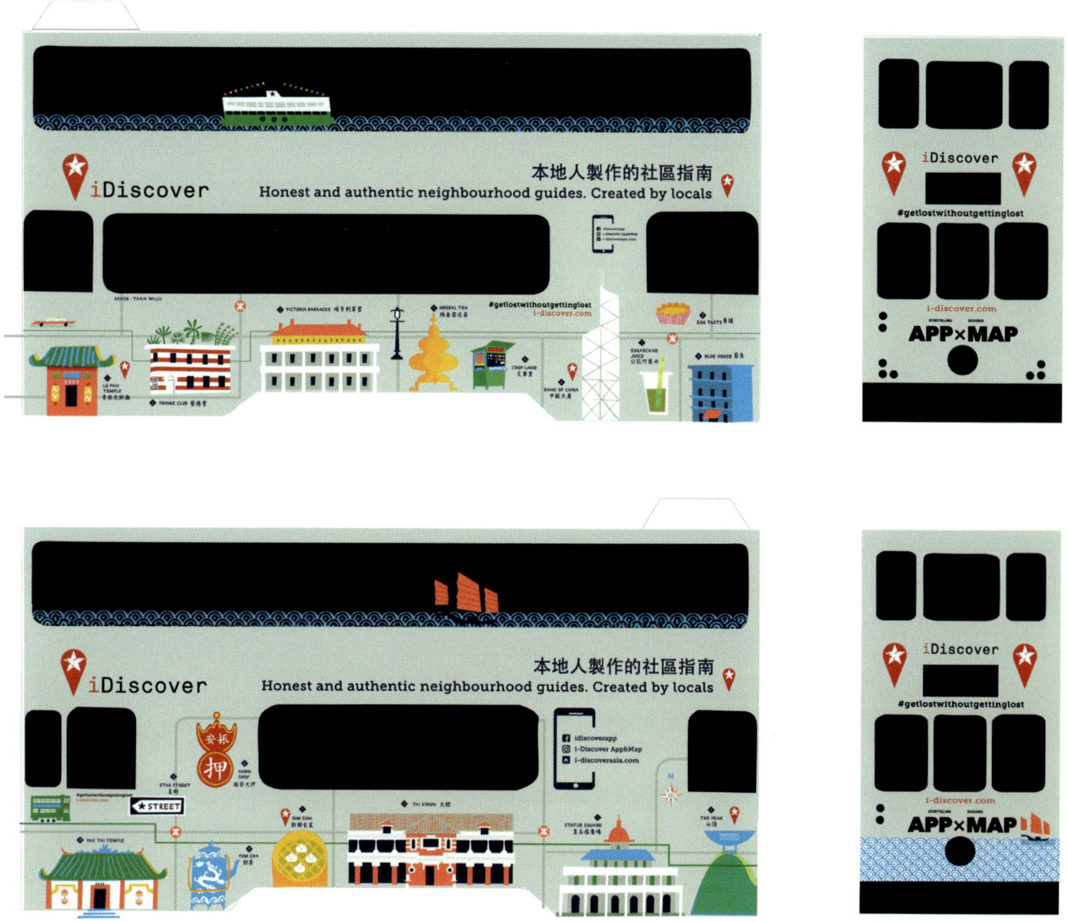

iDiscover
本地人製作的社區指南
Honest and authentic neighbourhood guides. Created by locals
#getlostwithoutgettinglost
i-discover.com
APP×MAP

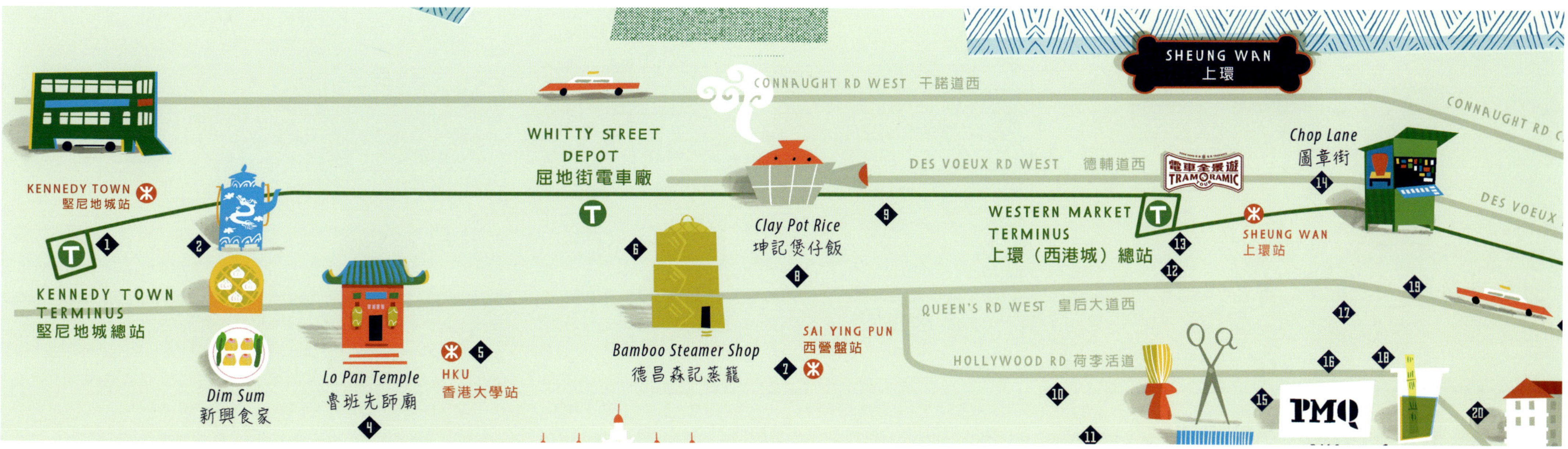

SHEUNG WAN
上環
CONNAUGHT RD WEST 干諾道西
CONNAUGHT RD
WHITTY STREET
DEPOT
屈地街電車廠
DES VOEUX RD WEST 德輔道西
DES VOEUX
TRAMORAMIC
Chop Lane
圖章街
KENNEDY TOWN
堅尼地城站
Clay Pot Rice
坤記煲仔飯
WESTERN MARKET
TERMINUS
上環（西港城）總站
SHEUNG WAN
上環站
KENNEDY TOWN
TERMINUS
堅尼地城總站
QUEEN'S RD WEST 皇后大道西
SAI YING PUN
西營盤站
HOLLYWOOD RD 荷李活道
Dim Sum
新興食家
Lo Pan Temple
魯班先師廟
HKU
香港大學站
Bamboo Steamer Shop
德昌森記蒸籠
PMQ

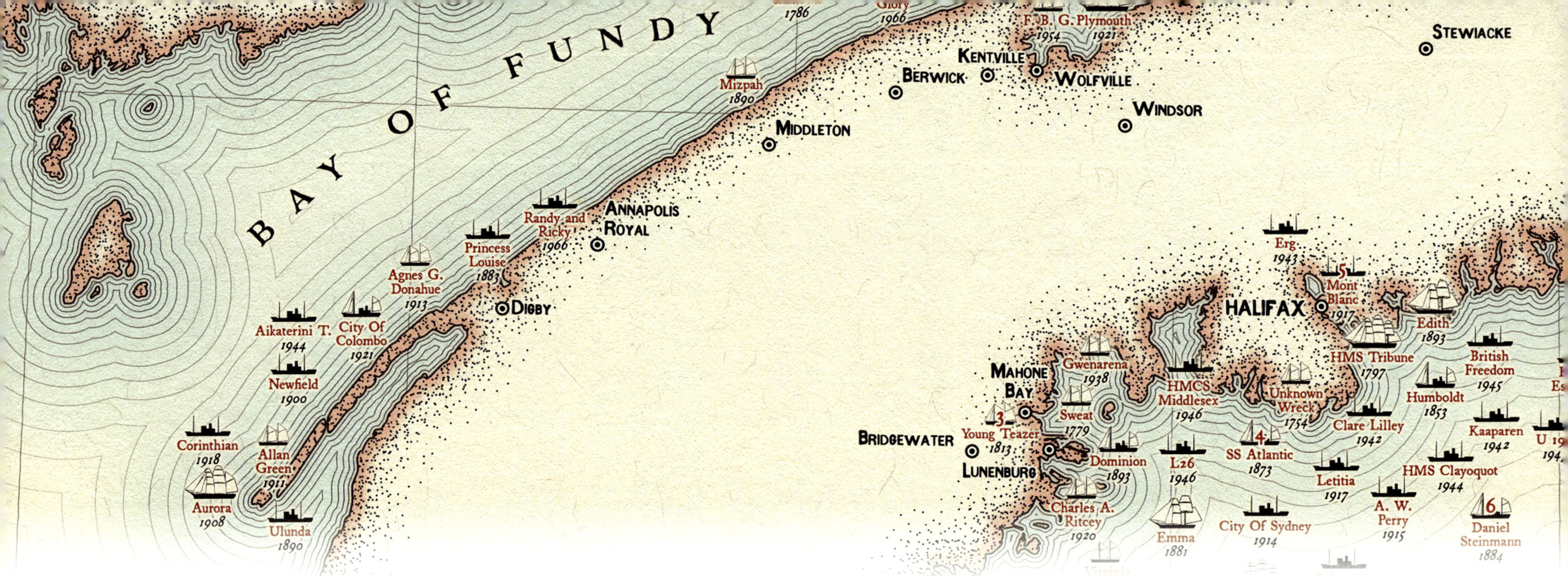

Shipwrecks of Nova Scotia »

Peter Atwood

Local history is one of my favorite subjects for mapping. I love having the chance to dive into the often disorganized, frequently confusing records to see if I can find an interesting geographic story to tell.

In this map for the Maritime Museum of the Atlantic, the museum's 5,000-entry shipwreck catalog served as a starting point. From there, I wanted to narrow it down to a list of significant wrecks spanning the province whose date, name, and location I could pinpoint. This turned out to be easier said than done. Only a handful of recent wrecks had their latitude and longitude recorded. The locations for the rest ranged from the vague ("near Halifax") to the ambiguous ("3 Miles SE of Big Island"; there are more than twenty islands of that name in Nova Scotia), to the unhelpful (217 wrecks "at sea"). Similarly, matching a specific ship to a wreck presented some challenges, with many names appearing multiple times. In some cases, these were simply popular names for ships (*Mary*, *Elizabeth*, and *Maggie* each appear more than ten times). In other cases, they were the result of a single, extremely accident-prone ship (the steamer *A. W. Perry* appears nine times).

With the help of local archives, newspaper records, and dive charts, I ended up with a list of about 120 wrecks to map. In some cases these wrecks had a significant human impact, either from a great loss of life, or by shaping maritime safety and lifesaving practices. Others were politically or militarily significant. Still others were the only wrecks that I could locate along a given section of coastline. I included these to show that almost every stretch of Nova Scotia's shores has its share of wrecks—even ones that have been largely forgotten.

For the design, the use of stippled coastlines, hatched shadows, and antique serif type laid over a parchment texture is meant to evoke the engraved maps of the 1800s, when maritime activity in Nova Scotia was at its peak. In a world of digital mapmaking, I always enjoy the opportunity to reinsert a bit of noise and randomness into my work.

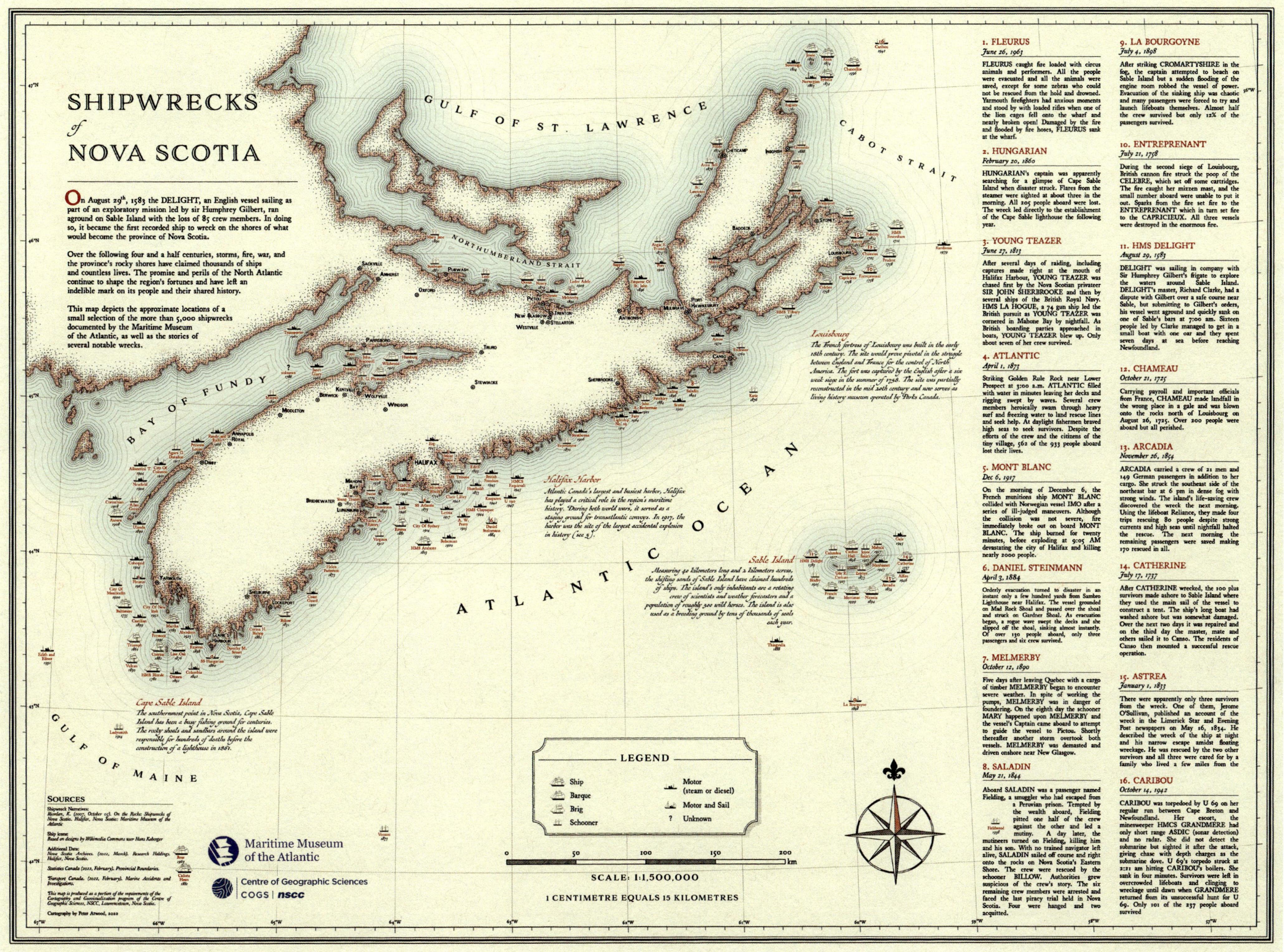
SHIPWRECKS
of
NOVA SCOTIA
On August 29th, 1583 the DELIGHT, an English vessel sailing as part of an exploratory mission led by sir Humphrey Gilbert, ran aground on Sable Island with the loss of 85 crew members. In doing so, it became the first recorded ship to wreck on the shores of what would become the province of Nova Scotia.
Over the following four and a half centuries, storms, fire, war, and the province's rocky shores have claimed thousands of ships and countless lives. The promise and perils of the North Atlantic continue to shape the region's fortunes and have left an indelible mark on its people and their shared history.
This map depicts the approximate locations of a small selection of the more than 5,000 shipwrecks documented by the Maritime Museum of the Atlantic, as well as the stories of several notable wrecks.
GULF OF ST. LAWRENCE
CABOT STRAIT
NORTHUMBERLAND STRAIT
BAY OF FUNDY
ATLANTIC OCEAN
GULF OF MAINE
Louisbourg
The French fortress of Louisbourg was built in the early 18th century. The site would prove pivotal in the struggle between England and France for the control of North America. The fort was captured by the English after a six week siege in the summer of 1758. The site was partially reconstructed in the mid 20th century and now serves as living history museum operated by Parks Canada.
Halifax Harbor
Atlantic Canada's largest and busiest harbor, Halifax has played a critical role in the region's maritime history. During both world wars, it served as a staging ground for transatlantic convoys. In 1917, the harbor was the site of the largest accidental explosion in history (see 5).
Sable Island
Measuring 40 kilometers long and 2 kilometers across, the shifting sands of Sable Island have claimed hundreds of ships. The island's only inhabitants are a rotating crew of scientists and weather forecasters and a population of roughly 500 wild horses. The island is also used as a breeding ground by tens of thousands of seals each year.
Cape Sable Island
The southernmost point in Nova Scotia, Cape Sable Island has been a busy fishing ground for centuries. The rocky shoals and sandbars around the island were responsible for hundreds of deaths before the construction of a lighthouse in 1861.
LEGEND
Ship
Barque
Brig
Schooner
Motor (steam or diesel)
Motor and Sail
? Unknown
0 50 100 150 200 km
SCALE: 1:1,500,000
1 CENTIMETRE EQUALS 15 KILOMETRES
SOURCES
Shipwreck Narratives:
Riordan, K. (2007, October 05). On the Rocks: Shipwrecks of Nova Scotia. Halifax, Nova Scotia: Maritime Museum of the Atlantic
Ship icons:
Based on designs by Wikimedia Commons user Hans Koberger
Additional Data:
Nova Scotia Archives. (2022, March). Research Holdings. Halifax, Nova Scotia.
Statistics Canada (2022, February). Provincial Boundaries.
Transport Canada. (2022, February). Marine Accidents and Investigations.
This map is produced as a portion of the requirements of the Cartography and Geovisualization program of the Centre of Geographic Sciences, NSCC, Lawrencetown, Nova Scotia.
Cartography by Peter Atwood, 2022
Maritime Museum of the Atlantic
Centre of Geographic Sciences
COGS | nscc
1. FLEURUS
June 26, 1963
FLEURUS caught fire loaded with circus animals and performers. All the people were evacuated and all the animals were saved, except for some zebras who could not be rescued from the hold and drowned. Yarmouth firefighters had anxious moments and stood by with loaded rifles when one of the lion cages fell onto the wharf and nearly broken open! Damaged by the fire and flooded by fire hoses, FLEURUS sank at the wharf.
2. HUNGARIAN
February 20, 1860
HUNGARIAN's captain was apparently searching for a glimpse of Cape Sable Island when disaster struck. Flares from the steamer were sighted at about three in the morning. All 205 people aboard were lost. The wreck led directly to the establishment of the Cape Sable lighthouse the following year.
3. YOUNG TEAZER
June 27, 1813
After several days of raiding, including captures made right at the mouth of Halifax Harbour, YOUNG TEAZER was chased first by the Nova Scotian privateer SIR JOHN SHERBROOKE and then by several ships of the British Royal Navy. HMS LA HOGUE, a 74 gun ship led the British pursuit as YOUNG TEAZER was cornered in Mahone Bay by nightfall. As British boarding parties approached in boats, YOUNG TEAZER blew up. Only about seven of her crew survived.
4. ATLANTIC
April 1, 1873
Striking Golden Rule Rock near Lower Prospect at 3:00 a.m. ATLANTIC filled with water in minutes leaving her decks and rigging swept by waves. Several crew members heroically swam through heavy surf and freezing water to land rescue lines and seek help. At daylight fishermen braved high seas to seek survivors. Despite the efforts of the crew and the citizens of the tiny village, 562 of the 933 people aboard lost their lives.
5. MONT BLANC
Dec 6, 1917
On the morning of December 6, the French munitions ship MONT BLANC collided with Norwegian vessel IMO after a series of ill-judged maneuvers. Although the collision was not severe, fire immediately broke out on board MONT BLANC. The ship burned for twenty minutes, before exploding at 9:05 AM devastating the city of Halifax and killing nearly 2000 people.
6. DANIEL STEINMANN
April 3, 1884
Orderly evacuation turned to disaster in an instant only a few hundred yards from Sambro Lighthouse near Halifax. The vessel grounded on Mad Rock Shoal and passed over the shoal and struck on Gardner Shoal. As evacuation began, a rogue wave swept the decks and she slipped off the shoal, sinking almost instantly. Of over 130 people aboard, only three passengers and six crew survived.
7. MELMERBY
October 12, 1890
Five days after leaving Quebec with a cargo of timber MELMERBY began to encounter severe weather. In spite of working the pumps, MELMERBY was in danger of foundering. On the eighth day the schooner MARY happened upon MELMERBY and the vessel's Captain came aboard to attempt to guide the vessel to Pictou. Shortly thereafter another storm overtook both vessels. MELMERBY was demasted and driven onshore near New Glasgow.
8. SALADIN
May 21, 1844
Aboard SALADIN was a passenger named Fielding, a smuggler who had escaped from a Peruvian prison. Tempted by the wealth aboard, Fielding pitted one half of the crew against the other and led a mutiny. A day later, the mutineers turned on Fielding, killing him and his son. With no trained navigator left alive, SALADIN sailed off course and right onto the rocks on Nova Scotia's Eastern Shore. The crew were rescued by the schooner BILLOW. Authorities grew suspicious of the crew's story. The six remaining crew members were arrested and faced the last piracy trial held in Nova Scotia. Four were hanged and two acquitted.
9. LA BOURGOYNE
July 4, 1898
After striking CROMARTYSHIRE in the fog, the captain attempted to beach on Sable Island but a sudden flooding of the engine room robbed the vessel of power. Evacuation of the sinking ship was chaotic and many passengers were forced to try and launch lifeboats themselves. Almost half the crew survived but only 12% of the passengers survived.
10. ENTREPRENANT
July 21, 1758
During the second siege of Louisbourg, British cannon fire struck the poop of the CELEBRE, which set off some cartridges. The fire caught her mizzen mast, and the small number aboard were unable to put it out. Sparks from the fire set fire to the ENTREPRENANT which in turn set fire to the CAPRICIEUX. All three vessels were destroyed in the enormous fire.
11. HMS DELIGHT
August 29, 1583
DELIGHT was sailing in company with Sir Humphrey Gilbert's frigate to explore the waters around Sable Island. DELIGHT's master, Richard Clarke, had a dispute with Gilbert over a safe course near Sable, but submitting to Gilbert's orders, his vessel went aground and quickly sank on one of Sable's bars at 7:00 am. Sixteen people led by Clarke managed to get in a small boat with one oar and they spent seven days at sea before reaching Newfoundland.
12. CHAMEAU
October 21, 1725
Carrying payroll and important officials from France, CHAMEAU made landfall in the wrong place in a gale and was blown onto the rocks north of Louisbourg on August 26, 1725. Over 200 people were aboard but all perished.
13. ARCADIA
November 26, 1854
ARCADIA carried a crew of 21 men and 149 German passengers in addition to her cargo. She struck the southeast side of the northeast bar at 6 pm in dense fog with strong winds. The island's life-saving crew discovered the wreck the next morning. Using the lifeboat Reliance, they made four trips rescuing 80 people despite strong currents and high seas until nightfall halted the rescue. The next morning the remaining passengers were saved making 170 rescued in all.
14. CATHERINE
July 17, 1737
After CATHERINE wrecked, the 100 plus survivors made ashore to Sable Island where they used the main sail of the vessel to construct a tent. The ship's long boat had washed ashore but was somewhat damaged. Over the next two days it was repaired and on the third day the master, mate and others sailed it to Canso. The residents of Canso then mounted a successful rescue operation.
15. ASTREA
January 1, 1833
There were apparently only three survivors from the wreck. One of them, Jerome O'Sullivan, published an account of the wreck in the Limerick Star and Evening Post newspapers on May 16, 1834. He described the wreck of the ship at night and his narrow escape amidst floating wreckage. He was rescued by the two other survivors and all three were cared for by a family who lived a few miles from the
16. CARIBOU
October 14, 1942
CARIBOU was torpedoed by U 69 on her regular run between Cape Breton and Newfoundland. Her escort, the minesweeper HMCS GRANDMERE had only short range ASDIC (sonar detection) and no radar. She did not detect the submarine but sighted it after the attack, giving chase with depth charges as the submarine dove. U 69's torpedo struck at 2:21 am hitting CARIBOU's boilers. She sank in four minutes. Survivors were left in overcrowded lifeboats and clinging to wreckage until dawn when GRANDMERE returned from its unsuccessful hunt for U 69. Only 101 of the 237 people aboard survived

« A Map of Home, Stitched in Found Fabrics

Eleanor Lutz

This map of the Upper East Side of Manhattan is a cozy combination of two things I love: cartography and quilting.

When I lived in New York City a few years ago, I spent a lot of time involved with the fabric art community. I was part of a quilter's guild that met up each month and worked on quilts together. And I also sorted fabrics for the Fabscrap warehouse in Brooklyn, a nonprofit that recycles fabric remnants from the fashion industry. With this quilt map, I wanted to make something cozy that represented everything I'd picked up in New York—both literally, in the form of fabric scraps, and metaphorically, as advice from my quilting guild.

All of the fabrics in this quilt are pieces that made their way to me as gifts from my neighborhood: lavender linen from a scrap exchange with my quilt guild, tiny samples of cotton shirting fabric from the Fabscrap sorting table, an old pair of my husband's gray jeans that ripped while he was biking through Manhattan.

The most complicated part of this map was making Central Park. I appliqued all of the lakes and baseball diamonds onto the ochre background, and embroidered all of the footpaths and major roads. This embroidery is probably my favorite addition, because you can use your fingers to trace your way through the park over the raised threads. To make the map design itself, I pasted together lots of screenshots from Google Maps and used that as a reference to calculate the sizes of city blocks and streets in Illustrator.

I don't live in New York City anymore. But this quilt still hangs in our living room as a cozy reminder of fun memories in the city, from our tiny wedding in Central Park to hot dogs and mulled wine at the Schaller & Weber hot dog counter in the freezing winter. A small appliqued heart marks the apartment block where we used to live.

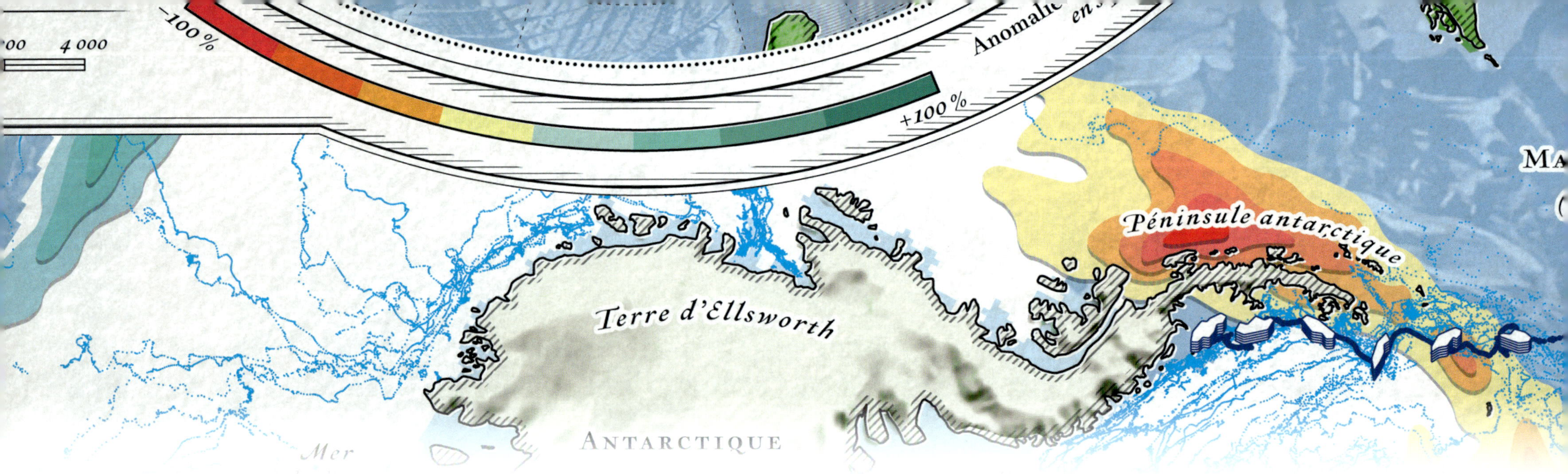

Une Mappemonde Craquelée »

Xemartin Laborde, Francesca Fattori, and Delphine Papin

Our polar ice caps are cracking, disintegrating, melting away. This is the landscape depicted by this world map, taken from the atlas *Mappemondes: Un voyage dans le temps pour raconter le monde contemporain*, published in 2023 by the French publisher Armand Colin, featuring fifty large world maps. This grand planisphere aligns with the tradition of old atlases that prominently featured physical and environmental geography. The two hemispheres are centered on the poles, illustrating the peril of the Arctic and Antarctic ice caps, which continue to diminish in both surface area and thickness. Similar to the layouts in old atlases where large maps were accompanied by smaller, zoomed-in maps, the focus is again on the poles, notably Antarctica, where melting leads to drifting icebergs whose movements have been tracked by satellite.

Reviving the vintage allure and charm of past world maps prevalent in twentieth-century atlases while representing current data on contemporary issues is the visual and aesthetic approach of this atlas. Creating surprise by playing with contrast and anachronism is the desired effect. Our grandparents' old atlases, with their comparative tables and choropleth and physiographic maps, were a treasure trove of information with abundant figures and meticulous detail. An elegant layout, engraved titling, relief shading, irregular patterns, arabesques, ornamental elements, compasses, and gridlines, along with lithographic imperfections like slight misalignments and tremulous lines, evoke a distinct design characteristic. The Sirenne typeface, with serifs reminiscent of fonts like Didot, Bodoni, or Garamond, enhances the typical retro ambiance of our attic-bound old books. As for the marine relief, I chose to project and colorize the hand-painted seabeds by Heinrich Berann from the 1970s.

This *Mappemondes* atlas originates from the #30DayMapChallenge, a social event in which participants follow a daily prompt and create a map per day over the course of a month. The idea of adhering to all the month's prompts, while maintaining the same design and riffing off the wordplay "30DayMappemondeChallenge," appealed to me, and I crafted thirty "world maps" addressing environmental issues (deforestation, rising sea levels, heat records, ocean and river pollution), as well as societal topics (gender and LGBT inequalities, press freedom, geopolitical conflicts). These are pressing, contemporary issues, coupled with a didactic and educational visual approach inspired by these historical "mappemondes."

UNE MAPPEMONDE CRAQUELÉE

Quand les Icebergs se Décrochent

— 22/41 —

Les signaux alarmants pour les glaces des pôles se multiplient. L'Arctique pourrait être privé de glace en été dès les années 2030, ont prévenu des chercheurs dans une étude, soit une décennie plus tôt que ce que prévoit le GIEC. Si aujourd'hui la banquise arctique a une superficie de 14 millions de kilomètres carrés, il ne devrait en rester pas plus d'un million à ce moment de l'année d'ici 10 ans.

La fonte de cette banquise, constituée d'eau douce, ne causerait pas d'élévation significative du niveau des océans, mais occasionnerait d'autres conséquences néfastes, en ne renvoyant plus les rayons du soleil, ce qui permet actuellement de maintenir l'Arctique à une température adaptée.

Même constat au pôle Sud. Alors que les niveaux de glace de l'Antarctique sont historiquement bas, des scientifiques ont exprimé leurs inquiétudes et craignent un effondrement global de la banquise.

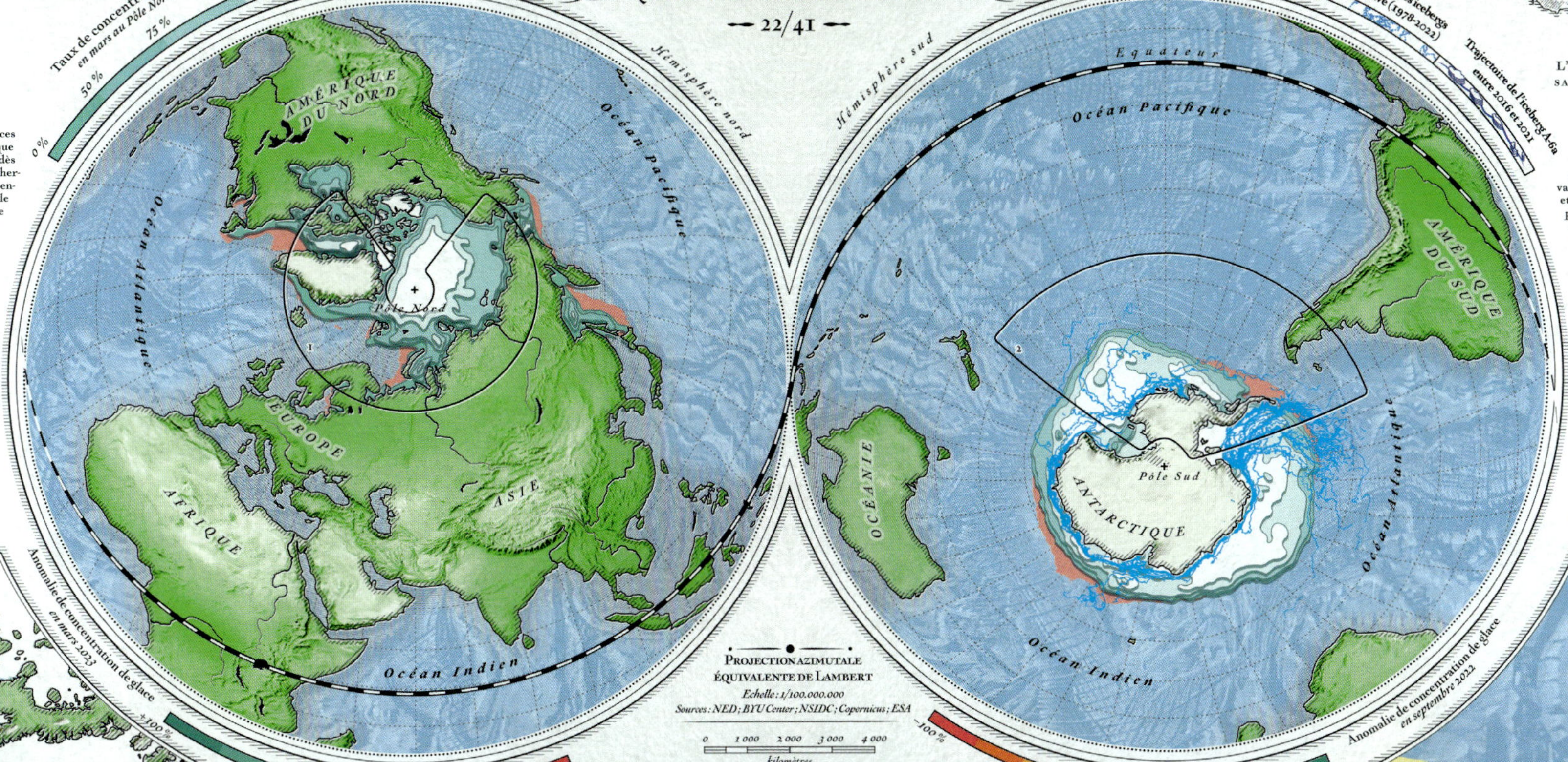

En février 2023, c'est-à-dire en plein été austral, cette dernière a atteint un niveau record d'atrophie, avec un déficit de glaces équivalent à la taille du Mexique. En juin et juillet 2023, la vague de chaleur la plus extrême a été enregistrée dans l'est du continent blanc, avec des relevés de températures excédant de 38,5 °C les normales enregistrées en 2022.

Symptôme de ce réchauffement, la banquise se fragilise et se craquelle, et des icebergs entrent en dérive. Ce fut le cas de l'iceberg A68, qui s'est détaché de la calotte glaciaire Larsen C, sur la péninsule Antarctique, en 2017. Il s'agit de l'un des plus gros icebergs jamais repérés, avec près de 5 800 kilomètres carrés, soit deux fois la taille du Luxembourg. Sa dérive, en direction des îles de Géorgie du Sud, dans l'Atlantique Sud, a été beaucoup médiatisée. Il s'est totalement dissous dans l'eau, relâchant quelque 152 milliards de tonnes d'eau douce.

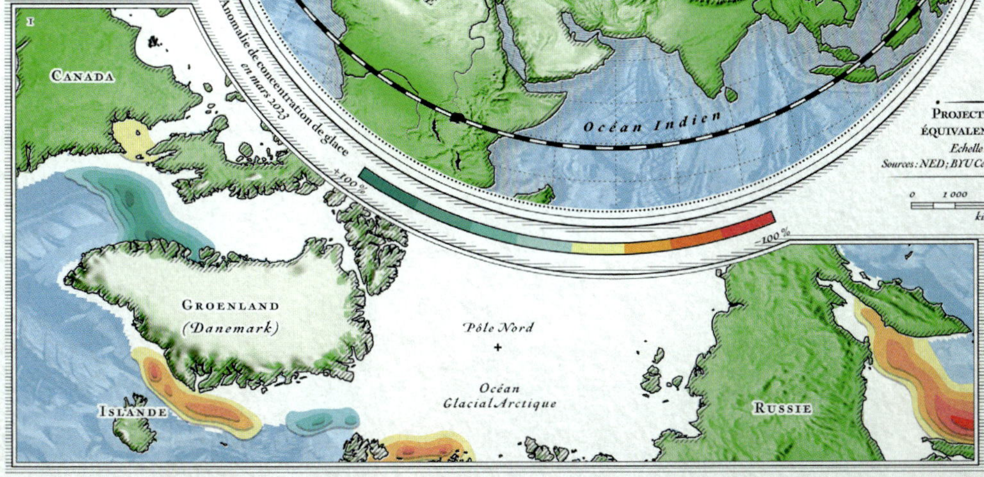

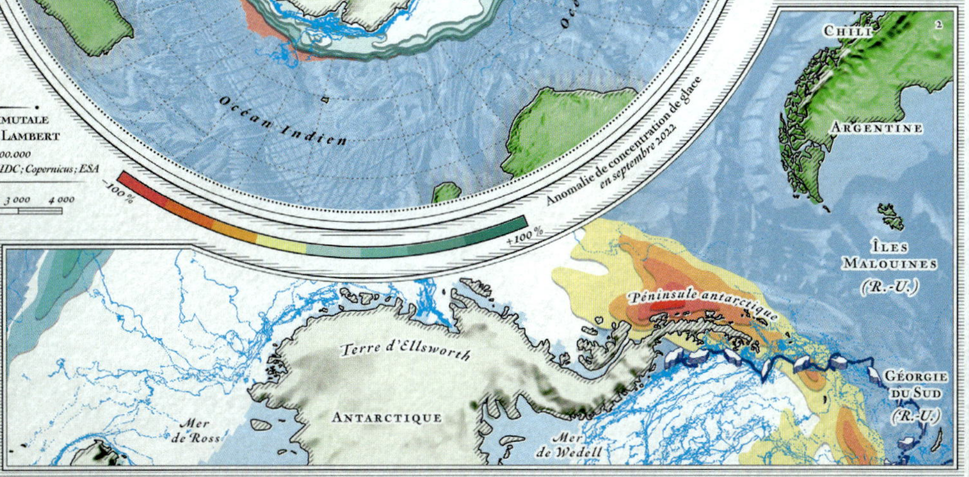

BEFORE YOU ARE HERE AN INDIGENOUS CARTOGRAPHY OF THE OHLONE BAY AREA
Cartographers Note: This map is a pastiche of projections and orientations from multiple perspectives. It has its own geography and scale, and collapses time into a single moment of reckoning: a cosmography rather than a cartography. Please view the map from any angle.
Summer Solstice 1578
Fall Equinox 1578
Winter Solstice 1578
Spring Equinox 1578
Presidio
Alcatraz
Yerba Buena Island
Angel Island
Brooks Island
Grizzly Peak

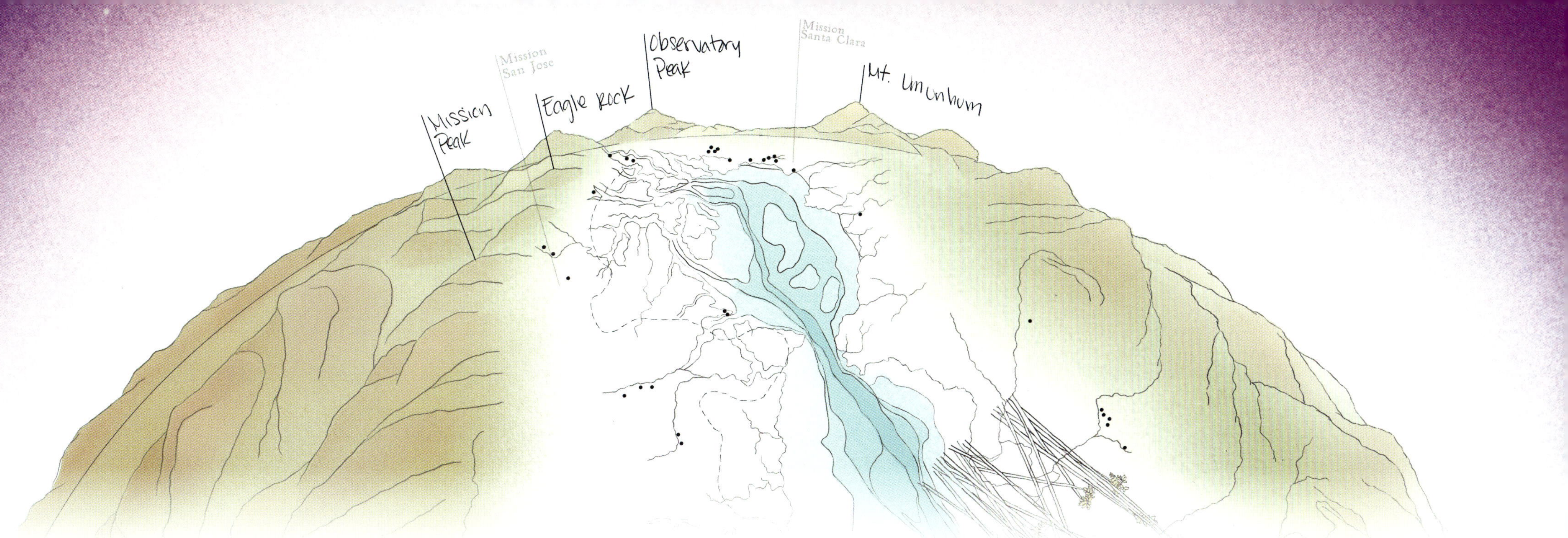

« Before You Are Here

Sogorea Te' Land Trust and Clancy Wilmott

"Before You Are Here" might be better described as a cosmography rather than a cartography. It forms part of an ongoing collaboration with the Sogorea Te' Land Trust, an urban Indigenous women-led land trust based in the San Francisco Bay Area. Sogorea Te' are Lisjan "Ohlone"–led, but intertribal, multicultural, and land-focused. Their aim is to "create opportunities for all people living in Lisjan territory to work together to re-envision the Bay Area community and what it means to live on Indigenous land," through land rematriation, cultural revitalization, and education programs.

This map challenges traditional cartographic paradigms by weaving together multiple spaces and times to depict the Bay Area as an ambiguous, relational, and deeply contested site, focusing on the *where*, rather than the *when*. It can be viewed from any angle, but cannot be viewed all at once—an effect achieved through the development of a bespoke stereographic projection, a series of irregularly sized rings centered on the first site rematriated by Sogorea Te', at Lisjan in East Oakland. To underscore the multiplicity of lives in the Bay Area, and to sidestep critiques of the cartographic "view from nowhere," the perspective is further formed by multiple "views from somewhere," looking up at the stars, down at the shoreline, and across the hills—a technique that was uncomputable and had to be developed using hand drafting techniques.

The unbroken line of the shoreline traces the original edges of the Bay, while the dotted line represents the contemporary shore; the contours of the Bay floor were calculated from soundings taken prior to the impacts of hydraulic mining in the nineteenth century. Celestial events, represented through the stars, embody the circularity of the seasons, as well as the way in which the stars have borne witness to the ongoingness of settler colonialism, winding anticlockwise from the view of the constellations over Tuyshtak on the winter solstice the year before contact with Western explorers. The labels are handwritten by Deja Gould, and work across both Chochenyo and settler names. The icons depicted are the return of the Chinook salmon and the monarch butterfly, the persistence of blue mussels and the coast live oaks, and the restorative roles played by Olympia oysters and tule.

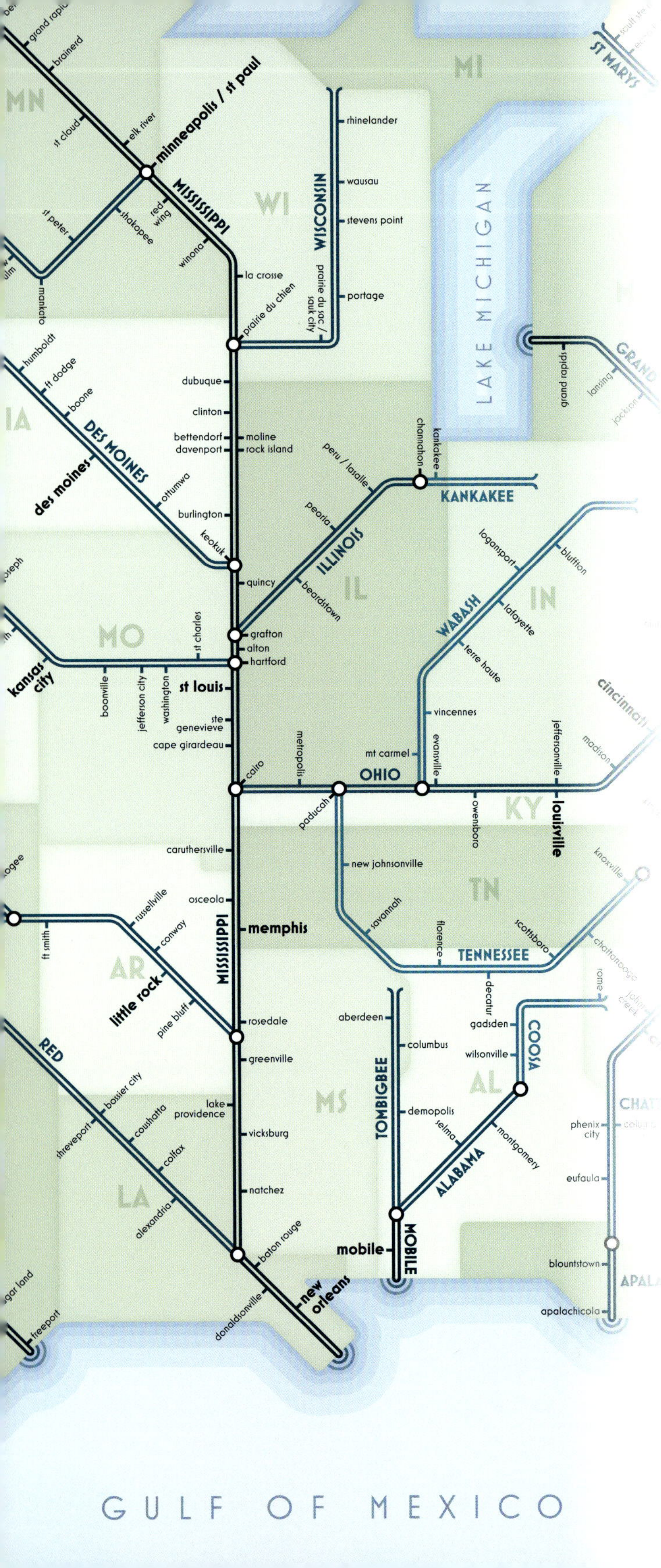

A Selection of North American Rivers »

Daniel P. Huffman, somethingaboutmaps

For more than a decade I've been making maps of rivers in a style that is meant to evoke the tropes of urban transit maps: straight lines that bend along a restricted set of angles, with nearby towns indicated as stops along the route. It's a fun visual theme to explore: its limitations force me to be creative, and the strong geometric emphasis of the river lines gives me a guide for how to craft other elements. For example, the type is set in Mostra Nuova, a geometric sans that's drawn from the Art Deco movement, while simple, perfect circles mark river junctions and mouths.

These maps also serve as a way for me to get comfortable with rather severe cartographic generalizations. I'm a detail-oriented perfectionist, and I've always been a little uncomfortable with the fact that I must simplify and abstract the underlying geography in order to make a map. Transit maps take this sort of generalization to a new level, and the process of making pieces like these forces me to practice a very loose style of representation. Here, any additional detail can potentially come at the expense of network clarity, so I am pushed into deciding what few pieces are critical for making a thumbnail sketch. It's a good exercise, one that often requires multiple iterations to get right.

I think this kind of transit design language makes sense for rivers because they have long been our highways. Their complex networks have historically driven why people live, farm, and trade in the places that they do. For most of my life I've lived near rivers and streams, and for a long time I never questioned where they came from or went, or even how they connected me to other cities. Just as a transit map sacrifices some spatial fidelity to disentangle rail lines and highlight connections, my maps set out to do the same for water flows. The process of making maps like this one has given me a deeper understanding and appreciation of the streams that surround me.

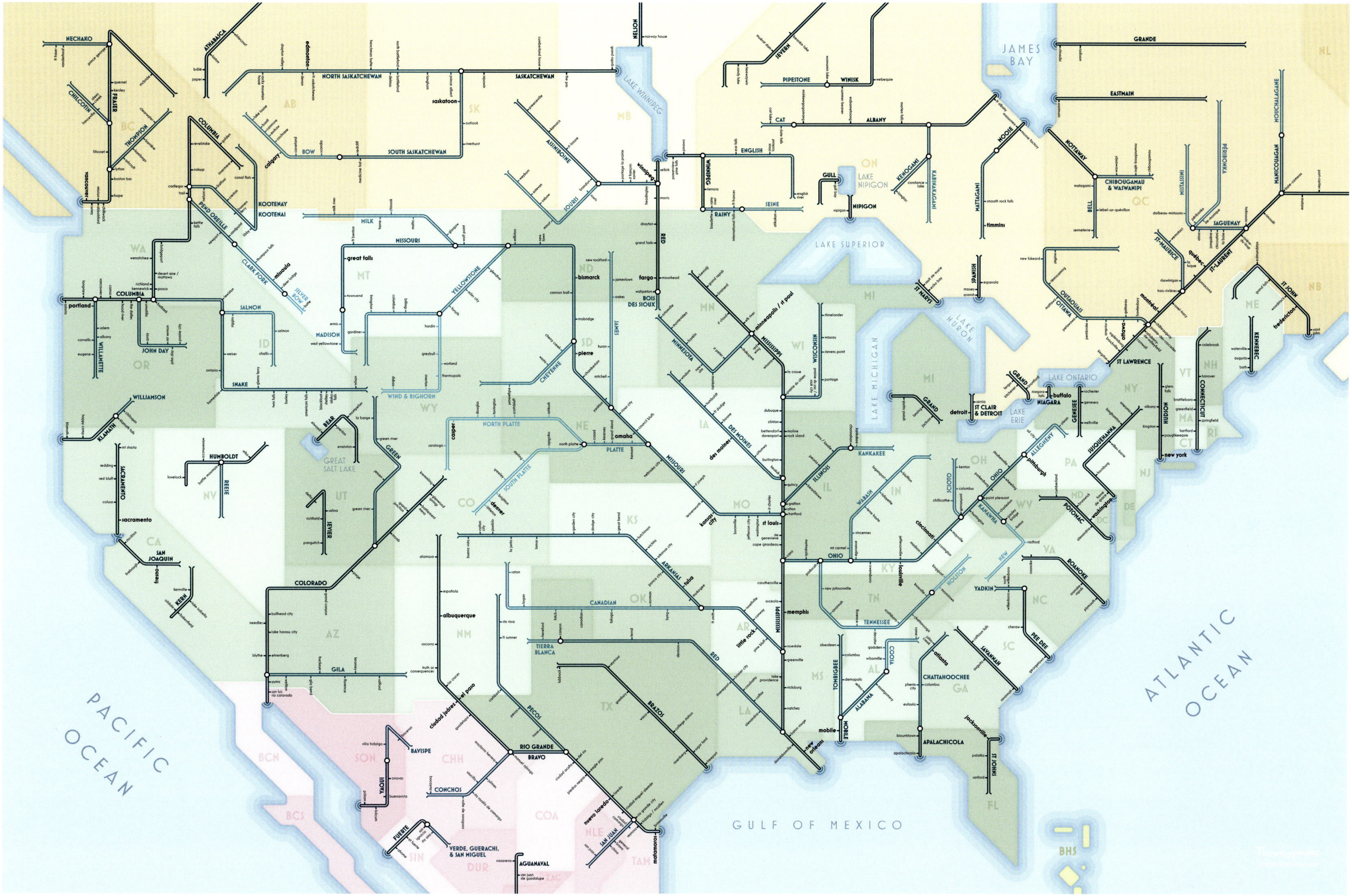
PACIFIC OCEAN
ATLANTIC OCEAN
GULF OF MEXICO
JAMES BAY
LAKE WINNIPEG
LAKE SUPERIOR
LAKE NIPIGON
LAKE MICHIGAN
LAKE HURON
LAKE ONTARIO
LAKE ERIE
GREAT SALT LAKE
BC
AB
SK
MB
ON
QC
NB
NL
WA
OR
CA
NV
ID
MT
WY
UT
AZ
CO
NM
ND
SD
NE
KS
OK
TX
MN
IA
MO
AR
LA
WI
IL
MI
IN
OH
KY
TN
MS
AL
GA
FL
SC
NC
VA
WV
PA
NY
ME
NH
VT
MA
CT
RI
NJ
DE
MD
BCN
BCS
SON
CHH
COA
NLE
TAM
SIN
DUR
ZAC
BHS
NECHAKO
FRASER
CHILCOTIN
THOMPSON
COLUMBIA
ATHABASCA
NORTH SASKATCHEWAN
SASKATCHEWAN
SOUTH SASKATCHEWAN
BOW
NELSON
KOOTENAY
KOOTENAI
PEND OREILLE
CLARK FORK
SILVER BOW
SALMON
SNAKE
JOHN DAY
WILLAMETTE
KLAMATH
WILLIAMSON
SACRAMENTO
SAN JOAQUIN
KERN
HUMBOLDT
REESE
BEAR
GREEN
SEVIER
COLORADO
GILA
MILK
MISSOURI
MADISON
YELLOWSTONE
WIND & BIGHORN
NORTH PLATTE
SOUTH PLATTE
PLATTE
CHEYENNE
JAMES
SOURIS
ASSINIBOINE
RED
BOIS DES SIOUX
WINNIPEG
RAINY
SEINE
ENGLISH
MINNESOTA
MISSISSIPPI
WISCONSIN
DES MOINES
ARKANSAS
CANADIAN
TIERRA BLANCA
BRAZOS
PECOS
RIO GRANDE
BRAVO
CONCHOS
YAQUI
BAVISPE
FUERTE
VERDE, GUERACHI, & SAN MIGUEL
AGUANAVAL
SAN JUAN
ILLINOIS
KANKAKEE
WABASH
OHIO
SCIOTO
KANAWHA
NEW
ALLEGHENY
TENNESSEE
HOLSTON
TOMBIGBEE
MOBILE
ALABAMA
COOSA
CHATTAHOOCHEE
APALACHICOLA
ST JOHNS
SAVANNAH
PEE DEE
YADKIN
ROANOKE
POTOMAC
SUSQUEHANNA
HUDSON
CONNECTICUT
GENESEE
NIAGARA
GRAND
ST CLAIR & DETROIT
ST LAWRENCE
OUTAOUAIS
OTTAWA
ST MARYS
SPANISH
MATTAGAMI
MOOSE
KENOGAMI
KABINAKAGAMI
ALBANY
CAT
PIPESTONE
WINISK
SEVERN
GULL
NIPIGON
NOTTAWAY
CHIBOUGAMAU & WASWANIPI
BELL
EASTMAIN
GRANDE
MISTASSINI
PERIBONKA
MANICOUAGAN
SAGUENAY
ST-MAURICE
ST-LAURENT
ST JOHN
KENNEBEC
edmonton
calgary
saskatoon
winnipeg
fargo
bismarck
pierre
omaha
kansas city
st louis
minneapolis / st paul
des moines
memphis
little rock
new orleans
mobile
atlanta
jacksonville
louisville
cincinnati
pittsburgh
washington
new york
buffalo
detroit
ottawa
montréal
québec
fredericton
timmins
portland
missoula
great falls
casper
denver
sacramento
fresno
albuquerque
el paso
ciudad juárez
nuevo laredo
matamoros
tulsa
vancouver

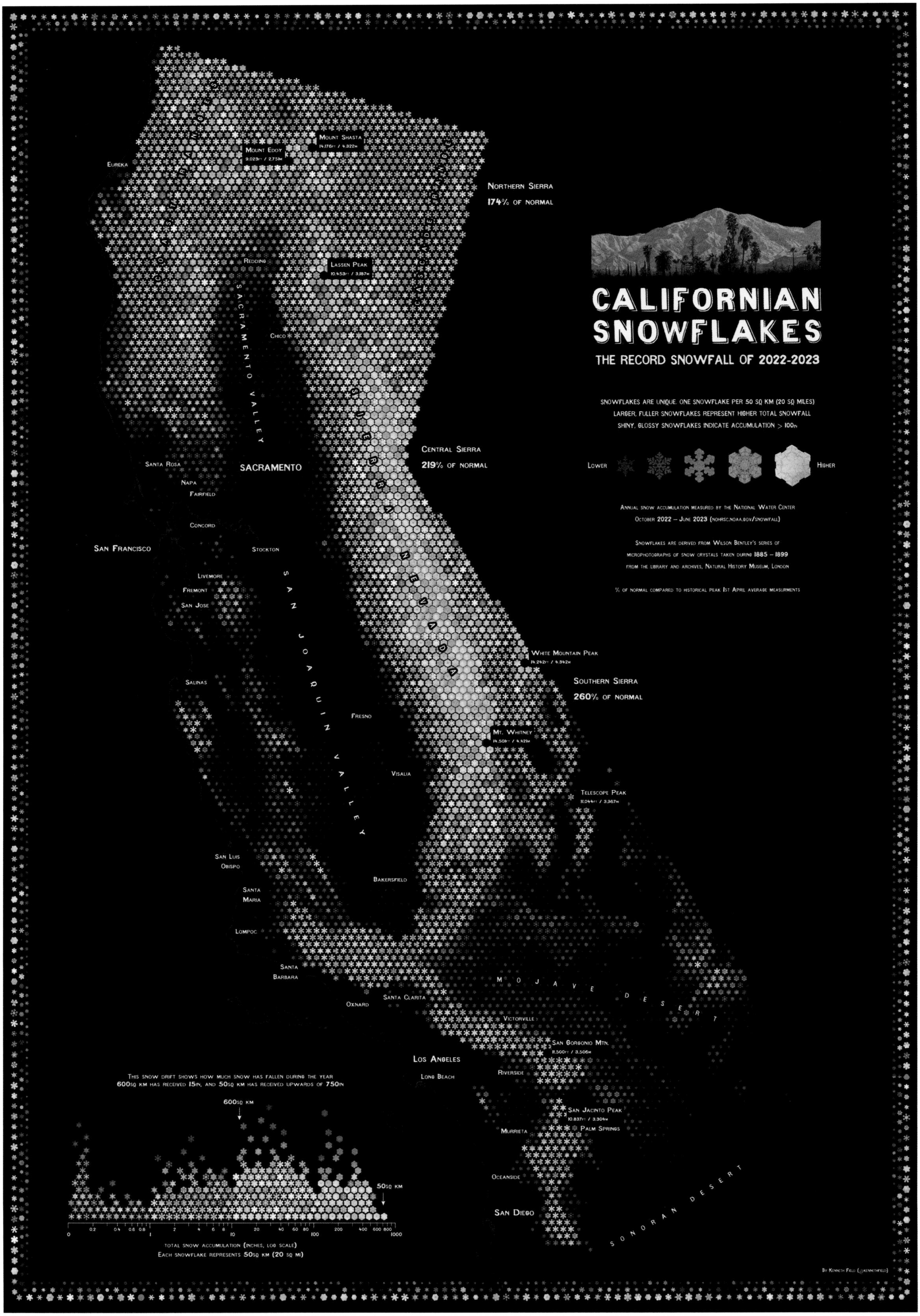
CALIFORNIAN
SNOWFLAKES
THE RECORD SNOWFALL OF 2022-2023
SNOWFLAKES ARE UNIQUE. ONE SNOWFLAKE PER 50 SQ KM (20 SQ MILES)
LARGER, FULLER SNOWFLAKES REPRESENT HIGHER TOTAL SNOWFALL
SHINY, GLOSSY SNOWFLAKES INDICATE ACCUMULATION > 100IN
Lower
Higher
Annual snow accumulation measured by the National Water Center
October 2022 – June 2023 (nohrsc.noaa.gov/snowfall)
Snowflakes are derived from Wilson Bentley's series of
microphotographs of snow crystals taken during 1885 – 1899
from the library and archives, Natural History Museum, London
% of normal compared to historical peak 1st April average measurments
Northern Sierra
174% of normal
Central Sierra
219% of normal
Southern Sierra
260% of normal
Mount Shasta
14,176ft / 4,322m
Mount Eddy
9,023ft / 2,751m
Lassen Peak
10,453ft / 3,187m
White Mountain Peak
14,242ft / 4,342m
Mt. Whitney
14,501ft / 4,421m
Telescope Peak
11,044ft / 3,367m
San Gorgonio Mtn.
11,500ft / 3,506m
San Jacinto Peak
10,837ft / 3,304m
Palm Springs
Eureka
Redding
Chico
SACRAMENTO VALLEY
SIERRA NEVADA
Santa Rosa
SACRAMENTO
Napa
Fairfield
Concord
San Francisco
Stockton
Livemore
Fremont
San Jose
SAN JOAQUIN VALLEY
Salinas
Fresno
Visalia
San Luis
Obispo
Santa
Maria
Bakersfield
Lompoc
Santa
Barbara
Oxnard
Santa Clarita
MOJAVE DESERT
Victorville
Los Angeles
Long Beach
Riverside
Murrieta
Oceanside
San Diego
SONORAN DESERT
This snow drift shows how much snow has fallen during the year
600sq km has received 15in, and 50sq km has received upwards of 750in
600sq km
50sq km
0 0.2 0.4 0.6 0.8 1 2 4 6 8 10 20 40 60 80 100 200 400 600 800 1000
total snow accumulation (inches, log scale)
Each snowflake represents 50sq km (20 sq mi)
By Kenneth Field (@kennethfield)

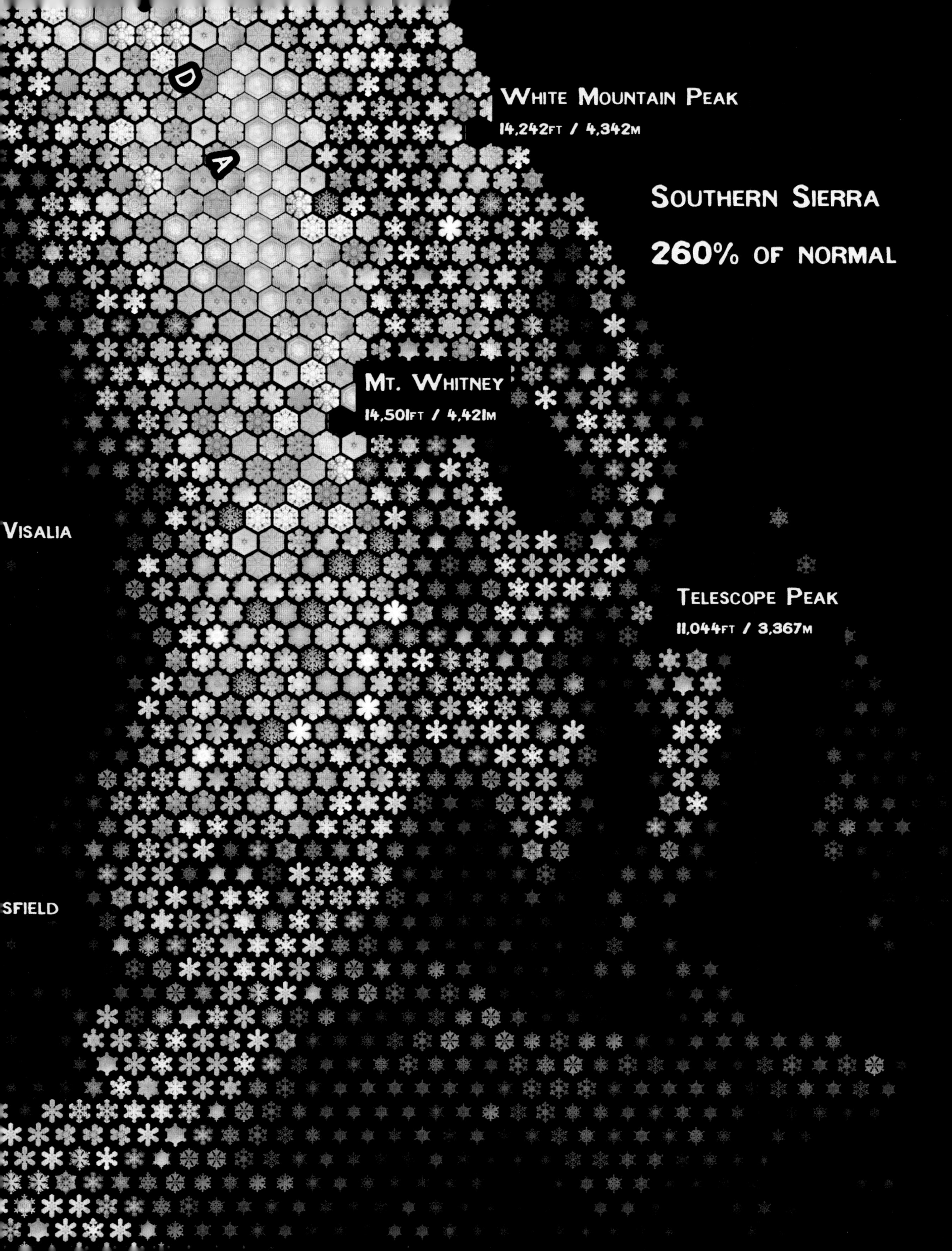

White Mountain Peak
14,242ft / 4,342m
Southern Sierra
260% of normal
Mt. Whitney
14,501ft / 4,421m
D
A
Visalia
Telescope Peak
11,044ft / 3,367m
sfield

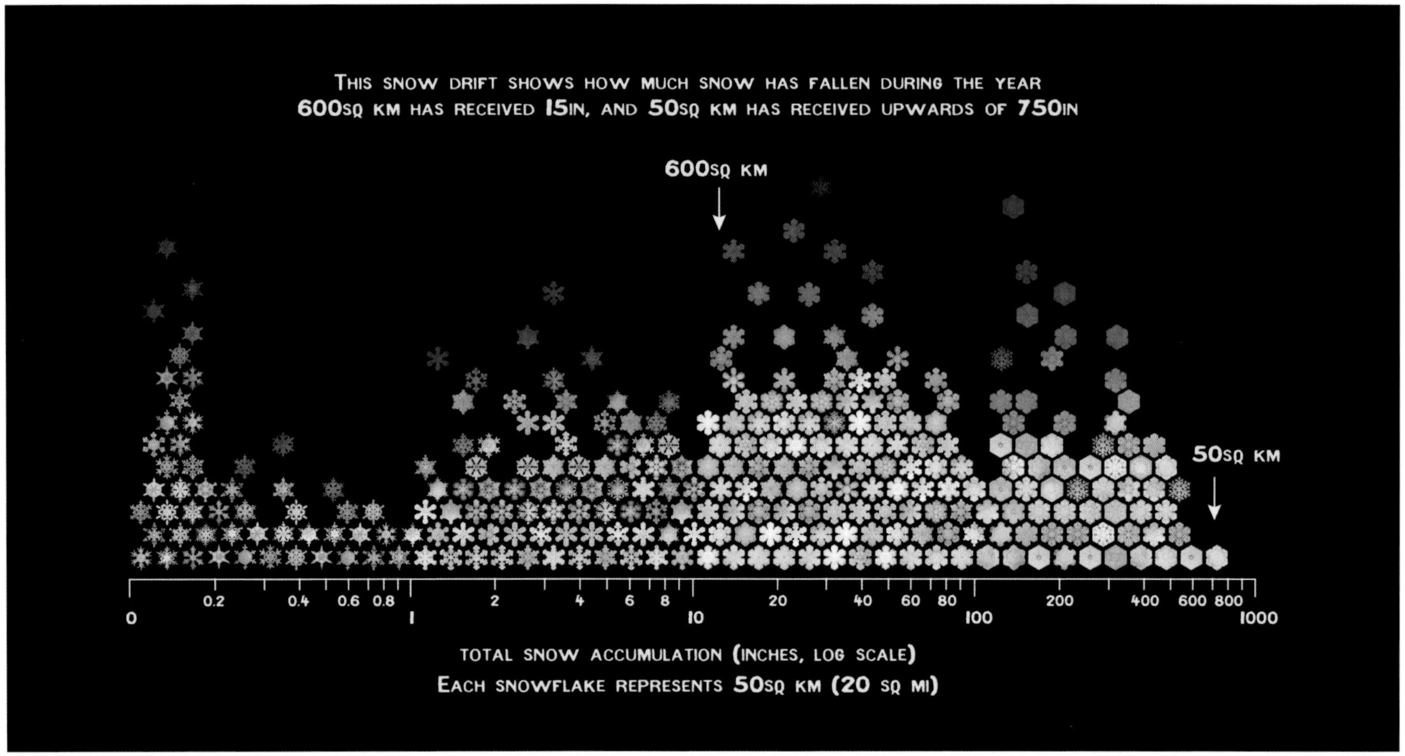

« Californian Snowflakes: The Record Snowfall of 2022–2023

Kenneth Field

Inspiration for maps can come from many different places, but usually there's a spark that sets the mind off down the path. In late 2022 I saw that the Natural History Museum (NHM) in London, England, had digitized a fabulous publication from 1899 by Wilson Bentley, otherwise known as "Snowflake Bentley," who was a meteorologist and photographer. He became famous for his microphotographs of individual snowflakes, and the books he published were collections of the photographs. Bentley's photographs go some way to illustrate that no two snowflakes are the same except that they all broadly exist in a hexagonal, six-sided form.

With record rain and snowfall impacting California in the early part of 2023, a slew of weather "maps" began to appear. Typically, they were full of rainbow colors, which is really not the most effective way of representing this phenomenon. (Different hues are better used for representing qualitative, not quantitative data because the human brain doesn't see hues as empirically linear.) Serendipity made it just the right time to create a hex-binned map of California's record snowfall using Bentley's microphotographs. This map illustrates snowfall using 50 sq km hex-bins that each show the average total accumulation. Higher total accumulation is shown using fuller, opaque snowflakes from Bentley's collection, while lower accumulation is shown with smaller, wispy snowflakes, modulated by size and transparency. This gives the impression of more to less across the map, with nearly 8,000 (almost all different) white snowflakes set across a black background.

On the lithograph, the snowflakes are printed using Pantone Spot metallic silver to give them a hint of a sparkle, suggestive of how snow glistens on a crisp winter's morning. A spot gloss varnish is used as a further visual variable that represents the areas that gained more than 100 inches (250 cm) total snow accumulation. The purpose here is to use an artistic form of map symbology where snowflakes have seemingly randomly settled on the map, coupled with innovative printing techniques, to enhance and complement data representation. And why not use real snow to make a map of snow!

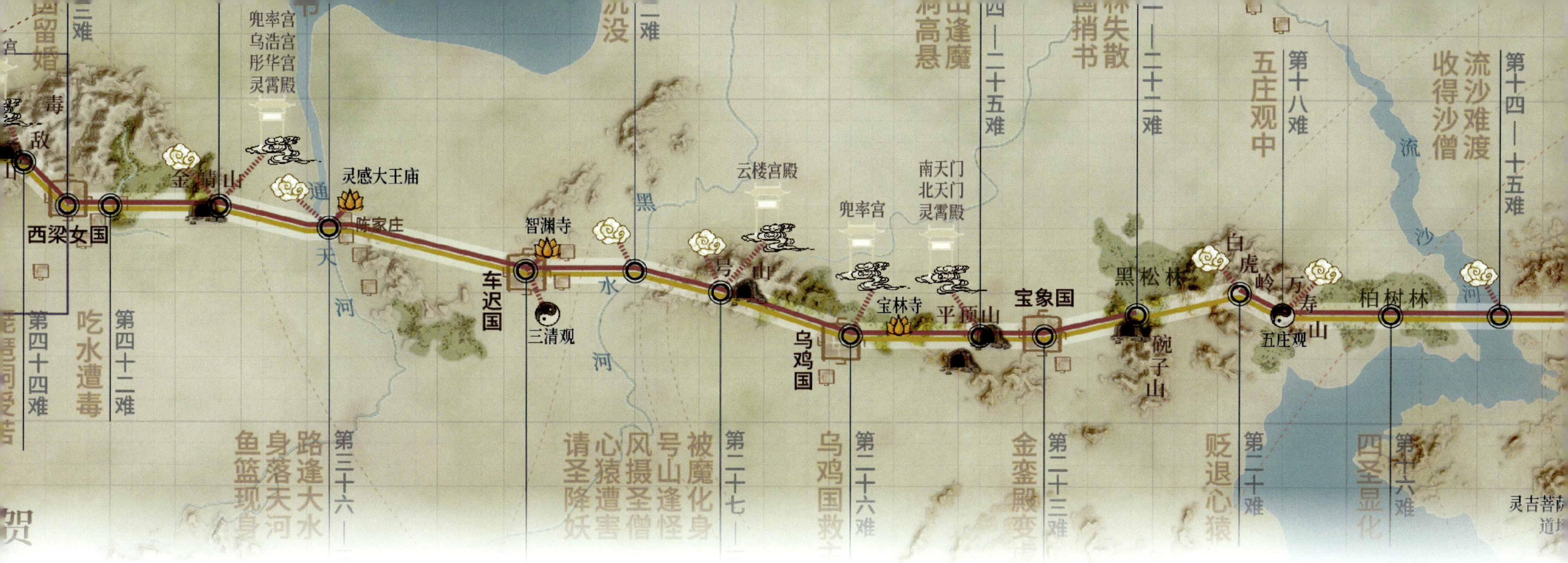

Exploratory Map of the Pilgrimage Route in *Journey to the West* »

Yue ZHANG, Jinnuo DUAN, and Xi TANG, Institute of Cartography, East China Normal University

Journey to the West (first published in 1592, but with deep roots in Chinese folklore and tradition) is the first long romantic novel about gods and demons in ancient China, and it exhibits many types of spaces and methods of travel through those spaces. This project attempts to present the spatial elements of the novel in one unprecedented map. To achieve this, we first shaped the fictional geographic space depicted in the novel, then organized and expressed the characters' locations and spatial movements in the storyline.

The space in *Journey to the West* is a unique blend of real and imaginary, posing a significant mapping challenge. We addressed this by referencing the real-world geography from the travels and literature of Buddhist Master Monk Xuanzang (seventh century CE) and integrating it with the imaginary space. This integration involved maintaining the continental plates and coastlines of China, India, and the Indochina Peninsula during the Tang Dynasty (618–907 CE), while incorporating the prevalent spatial cognition of the four major continents under the influence of Buddhism. We also established a measured grid system, using li as a distance unit (roughly a third of a mile), to redefine the spatial and temporal scale of the map, aligning the geographical distance of 108,000 *li* with the temporal distance of fourteen years.

Although the exact route taken during the story is ambiguous, we rearranged the geography into a westward orientation and created a route that roughly aligns with the journey of the novel. However, the spatial relationships of these routes are highly complex, with movements including flying, walking, escaping, and so on, which span heaven, ghostdom, and the mortal realm. We clarify these patterns with symbols showing landmarks, character types, and movement methods, to better articulate the narrative progression.

Furthermore, this map renders the geomorphic environment and the landscape according to the depiction of the corresponding plot. The resulting map seeks to provide fans of this classic Chinese novel a new method of exploring the incredible journey to the West.

西游記取经历程探索地图
EXPLORATORY MAP OF THE PILGRIMAGE ROUTE IN JOURNEY TO THE WEST
洞府
寺庙
道观
城郭、村落
筋斗云
主线路径
支线路径
灵山
长安
贞观二十二年
64500里
West
500 1500
(里)
灵山
雷音寺
天竺
灭法国
比丘国
狮驼国
黄花观
朱紫国
小雷音寺
木仙庵
金光寺
平阳之地
西梁女国
幽冥地府
西海龙宫
西
海
西
牛
贺
第八十难 凌云渡脱胎
第七十八难 天竺招婚
第七十三—七十五难 失落兵器 会庆钉钯 竹节山遭难
第七十一难 隐雾山遇魔
第六十六—六十八难 辨认真邪 松林救怪 僧房卧病
第六十一—六十四难 路阻狮驼 怪分三色 城里遇灾 请佛收魔
第五十九难 七情迷没
第五十六—五十八难 朱紫国行医 拯救疲癃 降妖取后
赛城扫塔 取宝救僧
第四十三难 西梁国留婚
第七十九难 铜台府监禁
第七十六—七十七难 玄英洞受苦 赶捉犀牛
第七十二难 凤仙郡求雨
第七十难 灭法国难行
第六十九难 无底洞遭困
第六十五难 比丘救子
第五十五难 稀柿同秽
第五十二难 棘林吟咏
第四十七—四十九难 路阻火焰山 求取芭蕉扇 收缚魔王
第四十五—四十六难 再贬心猿 难辨猕猴
第四十四难 琵琶洞受苦
吃水遭毒
灵山
平阳之地
观音菩萨
孙悟空
雷音寺
玉真观
六耳猕猴
幽冥地府
平阳之地
灵霄殿
花果山
第四十七难
沙悟净
六耳猕猴
杨氏村舍
第四十四难
孙悟空
歇脚处
琵琶洞
南山洞人家
观音菩萨
普陀山
灵山
幽冥地府
真假美猴王
True and False Monkey King
第四十五难 再贬心猿
第四十六难 难辨猕猴

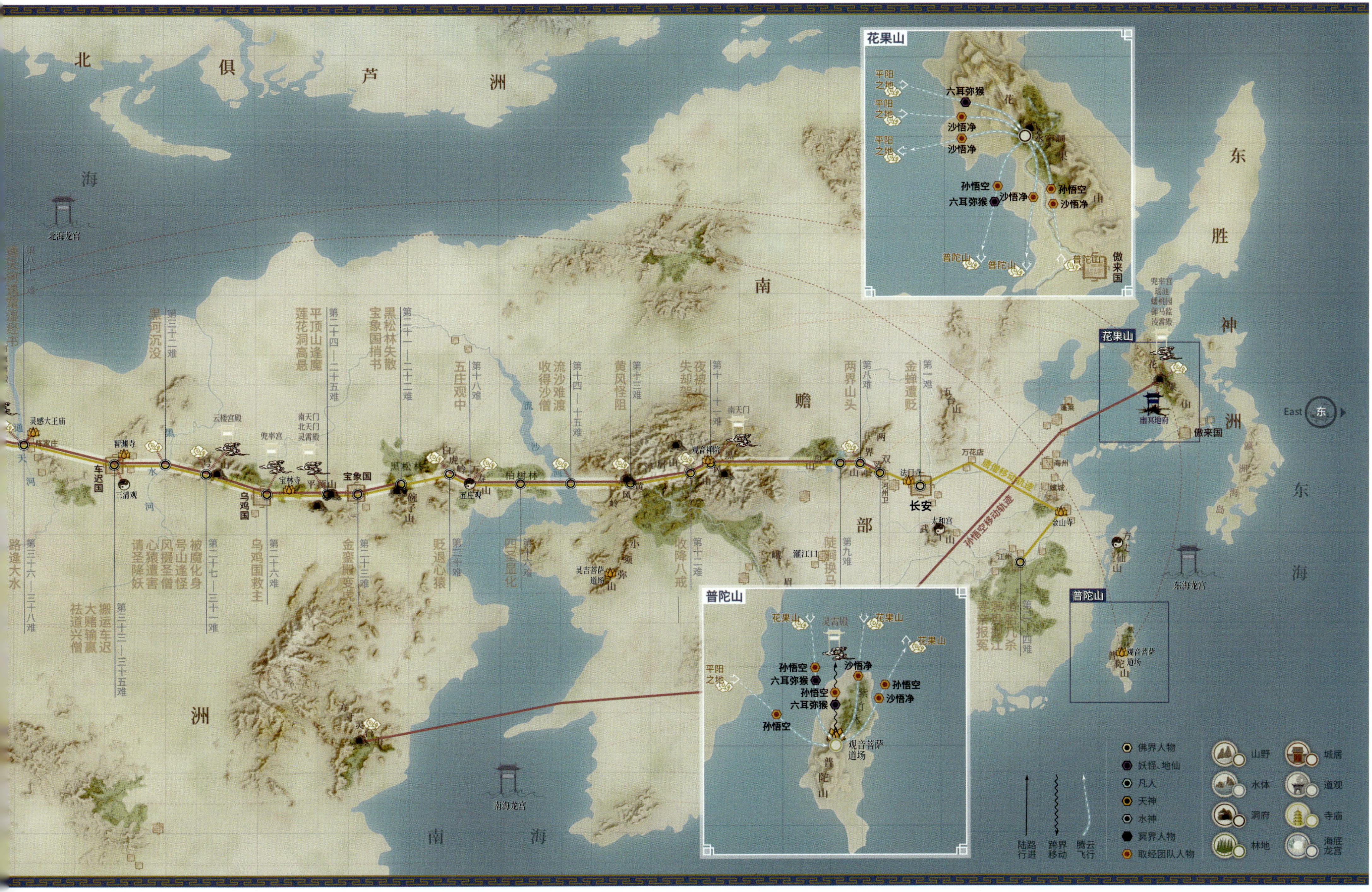

北
俱
芦
洲
海
北海龙宫
南
赡
部
洲
东
胜
神
洲
东
海
南
海
南海龙宫
东海龙宫
花果山
普陀山
长安
傲来国
车迟国
三清观
乌鸡国
宝象国
五庄观
法门寺
万花店
金山寺
大和宫
灌江口
唐僧移动轨迹
孙悟空移动轨迹
East 东
金蝉遭贬 第一难
两界山头 第八难
陡涧换马 第九难
夜被火烧 失却袈裟 第十—十一难
收降八戒 第十二难
黄风怪阻 第十三难
流沙难渡 收得沙僧 第十四—十五难
四圣显化 第十六难
五庄观中 第十八难
贬退心猿 第二十难
黑松林失散 宝象国捎书 第二十一—二十二难
金銮殿变虎 第二十三难
平顶山逢魔 莲花洞高悬 第二十四—二十五难
乌鸡国救主 第二十六难
被魔化身 号山逢怪 风摄圣僧 心猿遭害 请圣降妖 第二十七—三十一难
黑河沉没 第三十二难
搬运车迟 大赌输赢 祛道兴僧 第三十三—三十五难
路逢大水 第三十六—三十八难
花果山
平阳之地
六耳弥猴
沙悟净
孙悟空
普陀山
傲来国
普陀山
花果山
灵霄殿
孙悟空
六耳弥猴
沙悟净
平阳之地
观音菩萨道场
佛界人物
妖怪、地仙
凡人
天神
水神
冥界人物
取经团队人物
陆路行进
跨界移动
腾云飞行
山野
水体
洞府
林地
城居
道观
寺庙
海底龙宫

Spain's Muslim Era »

Rosemary P. Wardley, Fernando G. Baptista, Patricia Healy, and Eve Conant, National Geographic

"Spain's Muslim Era" was created as part of a larger graphics package, in collaboration with artist Fernando G. Baptista and researcher Patty Healy, to accompany a story about the Alhambra Palace in Granada, Spain. While Fernando crafted illustrations to give readers a detailed look at the buildings of the fortress complex, I (Rosemary Wardley) created these maps and the timeline to give context to the eight centuries in which Muslims ruled the Iberian Peninsula.

In researching this topic, I was struck by how quickly Islam spread to Europe, and the lasting impact of Islamic rule. However, this period is somewhat misleadingly referred to as the Reconquista, or the reconquest of Christian Spain from Muslim rulers. That name minimizes the golden age of scholarship and tolerant rule in Al Andalus—where Christians, Jews, and Muslims lived together, sharing their cultures. The map aims to highlight the lasting influence on culture, rather than focus solely on political rule or battles won and lost.

In order to capture this broad history, each piece of the map had a purpose, with information encoded into every part of the page. The main map details the shifting balance of power and highlights the influence of Islamic architecture across Spain. The timeline acts as a key to explain the complex rule of successive Muslim dynasties. The perspective map gives the reader a view of the Alhambra within the city of Granada, highlighting the physical location and the impressive hydraulic water systems constructed to feed the palace.

I worked with Fernando to add artistic design elements, such as the building icons and architectural frames, to the spread to connect the maps and graphics. One of my favorite elements is the placement of the building silhouettes on the timeline. While looking at a draft of this project, a friend suggested showing when each building was constructed. Sometimes as cartographers we get entangled in the minutia of a project. I've found that getting feedback from friends (and sometimes from my children) serves as a compass to help guide me back to the focus and purpose of my work.

SPAIN'S MUSLIM ERA

In the eighth century, an army of Arab Muslims crossed into the Iberian Peninsula, ousting Germanic Visigoths. Known as Al Andalus, their new land faced constant pressures yet became a hub of arts, education, and science for centuries.

BRIDGE TO EUROPE

Islam spreads from its birthplace on the Arabian Peninsula, in 632, to North Africa and Europe along trade routes. Defeat in France ends the advance.

HILLTOP FORTRESS

Water is central to construction, daily life, and religious ablutions in the Alhambra. The fortress's position high on Sabika Hill makes getting it there a challenge.

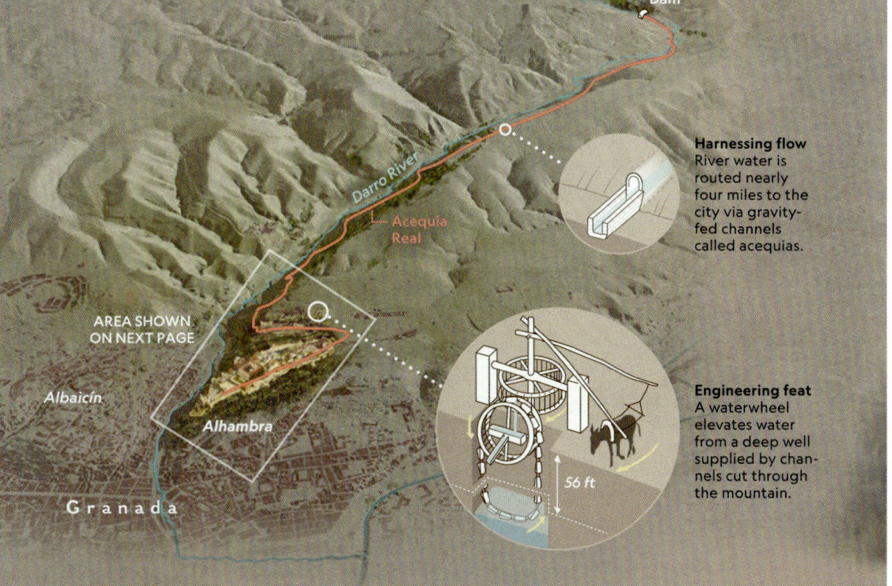

AREA SHOWN ON NEXT PAGE

AGE OF AL ANDALUS

First Crossing at Gibraltar
In 711, general Tariq ibn Ziyad sails a Muslim army toward the rock that would later bear his name.

1 Umayyad Dynasty
A break from the Damascus caliphate leads to the Iberian Peninsula's first independent Muslim state.

2 Almoravid Dynasty
North African Berber tribes fortify Muslim cities and strongholds against Christian advances.

3 Almohad Dynasty
New leaders move the capital from Córdoba to Seville, closer to the ocean for better trade.

4 Nasrid Dynasty
In a final power split, the last Muslim dynasty founds the Kingdom of Granada but later falls to Christian forces.

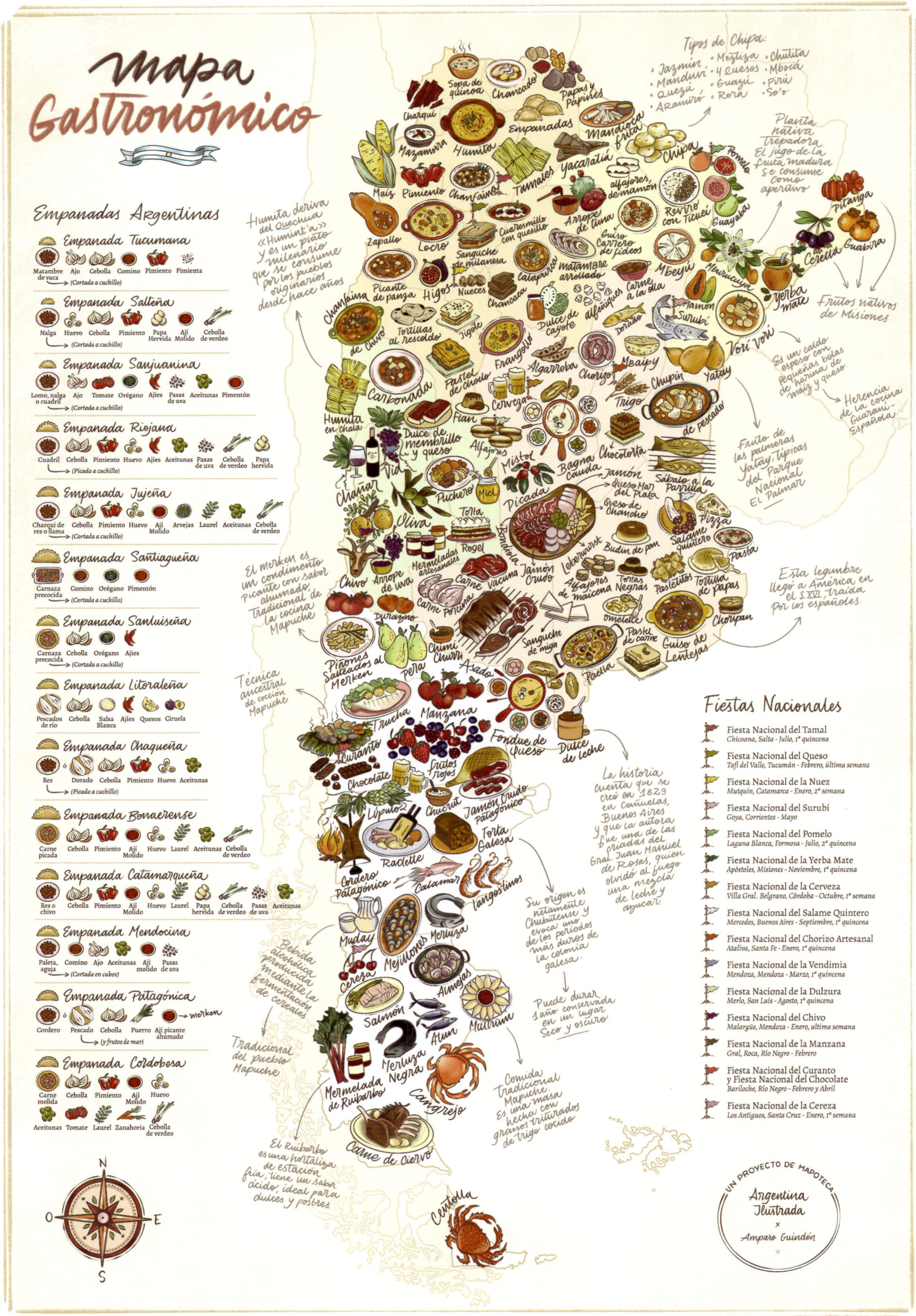

Mapa Gastronómico
Empanadas Argentinas
Empanada Tucumana
Matambre de vaca (Cortada a cuchillo) Ajo Cebolla Comino Pimiento Pimienta
Empanada Salteña
Nalga (Cortada a cuchillo) Huevo Cebolla Pimiento Papa Hervida Ají Molido Cebolla de verdeo
Empanada Sanjuanina
Lomo, nalga o cuadril (Cortada a cuchillo) Ajo Tomate Orégano Ajíes Pasas de uva Aceitunas Pimentón
Empanada Riojana
Cuadril (Picada a cuchillo) Cebolla Pimiento Huevo Ajíes Aceitunas Pasas de uva Cebolla de verdeo Papa hervida
Empanada Jujeña
Charqui de res o llama (Cortada a cuchillo) Cebolla Pimiento Huevo Ají Molido Arvejas Laurel Aceitunas Cebolla de verdeo
Empanada Santiagueña
Carnaza precocida (Cortada a cuchillo) Comino Orégano Pimentón
Empanada Sanluiseña
Carnaza precocida (Cortada a cuchillo) Cebolla Orégano Ajíes
Empanada Litoraleña
Pescados de río Cebolla Salsa Blanca Ajíes Quesos Ciruela
Empanada Chaqueña
Res ó Dorado (Picada a cuchillo) Cebolla Pimiento Huevo Aceitunas
Empanada Bonaerense
Carne picada Cebolla Pimiento Ají Molido Huevo Laurel Aceitunas Cebolla de verdeo
Empanada Catamarqueña
Res o chivo Cebolla Pimiento Ají Molido Huevo Laurel Papa hervida Cebolla de verdeo Pasas de uva Aceitunas
Empanada Mendocina
Paleta, aguja (Cortada en cubos) Comino Ajo Aceitunas Ají molido Pasas de uva
Empanada Patagónica
Cordero ó Pescado (y frutos de mar) Cebolla Puerro Ají picante ahumado → merken
Empanada Cordobesa
Carne molida Cebolla Pimiento Ají Molido Huevo Aceitunas Tomate Laurel Zanahoria Cebolla de verdeo
Tipos de Chipa:
· Jazmín · Mestiza · Chutita
· Manduví · 4 Quesos · Mbocá
· Quesú · Guazú · Pirú
· Aramiró · Rora · So'o
Planta nativa trepadora. El jugo de la fruta madura se consume como aperitivo
Frutos nativos de Misiones
Se un caldo espeso con pequeñas bolas de harina de maíz y queso.
Herencia de la cocina Guaraní-Española
Fruto de las palmeras Yatay, típicas del Parque Nacional El Palmar
Esta legumbre llegó a América en el S. XVI, traída por los españoles.
Humita deriva del Quechua «Humint'a» y es un plato milenario que se consume por los pueblos originarios desde hace años
El merken es un condimento picante con sabor ahumado. Tradicional de la cocina Mapuche
Técnica ancestral de cocción Mapuche
La historia cuenta que se creó en 1829 en Cañuelas, Buenos Aires y que la autora fue una de las criadas del Gral. Juan Manuel de Rosas, quien olvidó al fuego una mezcla de leche y azúcar.
Su origen es netamente Chubutense y evoca uno de los periodos más duros de la colonia galesa.
Puede durar 1 año conservada en un lugar seco y oscuro
Bebida alcohólica producida mediante la fermentación de cereales
Tradicional del pueblo Mapuche
Comida tradicional Mapuche. Es una masa hecha con granos triturados de trigo cocido
El Ruibarbo es una hortaliza de estación fría, tiene un sabor ácido, ideal para dulces y postres.
Sopa de quinoa
Chancado
Papas y Papines
Charqui
Empanadas
Mandioca frita
Chipa
Pomelo
Mazamorra
Humita
Tamales
Yacaratiá
Maíz
Pimiento
Chanfaina
Alfajores de mamón
Revirú con Ticuei
Guayaba
Pitanga
Cerella
Guabira
Arrope de tuna
Cuaresmillo con quesillo
Zapallo
Locro
Guiso Carrero de fideos
Mbeyú
Mburucuyá
Sanguche de milanesa
Yerba mate
Calapurca
Matambre arrollado
Picante de panza
Higos
Nueces
Chancaca
Carne a la olla
Mamón
Surubí
Dulce de cayote
Alfeñiques
Dorado
Vorí vorí
Tortillas al rescoldo
Jigote
Frangollo
Algarroba
Chorizo
Mbaipy
Chipín
Yatay
Pastel de choclo
Cerveza
Trigo
Carbonada
Flan
de pescado
Humita en chala
Dulce de membrillo y queso
Alfajores
Chocotorta
Vid
Místol
Bagna cauda
Jamón
Sábalo a la parrilla
Chañar
Pucheró
Miel
Picada
Queso Mar del Plata
Queso de Chancho
Oliva
Torta Rogel
Bondiola
Budín de pan
Salame quintero
Pizza
Mermeladas artesanales
Jamón Crudo
Leberwurst
Pastelitos
Pasta
Chivo
Arrope de uva
Carne Vacuna
Alfajores de maicena
Tortas Negras
Tortilla de papas
Carne Porcina
Omelette
Choripán
Durazno
Sanguche de miga
Pastel de carne
Guiso de Lentejas
Piñones Salteados al Merken
Pera
Chimi Churri
Asado
Paella
Trucha
Manzana
Fondue de Queso
Dulce de leche
Curanto
Frutos rojos
Chocolate
Lúpulo
Chucrut
Jamón Crudo patagónico
Torta Galesa
Raclette
Cordero Patagónico
Calamar
Langostinos
Muday
Merluza
Mejillones
Cereza
Almejas
Salmón
Mutrum
Atún
Merluza Negra
Mermelada de Ruibarbo
Cangrejo
Carne de Ciervo
Centolla
Fiestas Nacionales
Fiesta Nacional del Tamal
Chicoana, Salta - Julio, 1ª quincena
Fiesta Nacional del Queso
Tafí del Valle, Tucumán - Febrero, última semana
Fiesta Nacional de la Nuez
Mutquín, Catamarca - Enero, 2ª semana
Fiesta Nacional del Surubí
Goya, Corrientes - Mayo
Fiesta Nacional del Pomelo
Laguna Blanca, Formosa - Julio, 2ª quincena
Fiesta Nacional de la Yerba Mate
Apóstoles, Misiones - Noviembre, 1ª quincena
Fiesta Nacional de la Cerveza
Villa Gral. Belgrano, Córdoba - Octubre, 1ª semana
Fiesta Nacional del Salame Quintero
Mercedes, Buenos Aires - Septiembre, 1ª quincena
Fiesta Nacional del Chorizo Artesanal
Ataliva, Santa Fe - Enero, 1ª quincena
Fiesta Nacional de la Vendimia
Mendoza, Mendoza - Marzo, 1ª quincena
Fiesta Nacional de la Dulzura
Merlo, San Luis - Agosto, 1ª quincena
Fiesta Nacional del Chivo
Malargüe, Mendoza - Enero, última semana
Fiesta Nacional de la Manzana
Gral. Roca, Río Negro - Febrero
Fiesta Nacional del Curanto y Fiesta Nacional del Chocolate
Bariloche, Río Negro - Febrero y Abril
Fiesta Nacional de la Cereza
Los Antiguos, Santa Cruz - Enero, 1ª semana
N
O
E
S
UN PROYECTO DE MAPOTECA
Argentina Ilustrada
×
Amparo Guindón

« Argentina Gastronómica

Amparo Guindón, Mapoteca

Argentine cuisine is a blend of cultures—from the Indigenous peoples to the Mediterranean influences brought during the colonial period, later complemented with the influx of European and Middle Eastern migrants. "Argentina Gastronómica" offers a historical and geographical perspective on Argentine gastronomy. This map exemplifies how a story can be told without the need for lengthy explanatory text. With more than 100 illustrations meticulously crafted by hand through digital drawing techniques, it narrates a culinary journey by showcasing how food varies across the country.

This is not just a map; it is also an infographic that delves into the gastronomical festivities integral to the culture of each region. It explores fourteen distinct types of meat empanadas, each representing a province, and shares intriguing facts about many other dishes. This map serves as a testament to the fact that Argentine cuisine is influenced not solely by geography and available ingredients, but also by the diverse peoples who have contributed to its culinary heritage.

Before beginning the illustration process, it was necessary to conduct in-depth research on the typical dishes and ingredients of each province in order to select the most representative ones. Since Argentine cuisine is so varied and extensive, some dishes inevitably had to be left out, but we managed to reflect a part of each region's culinary identity with our representative selections. Once the dishes were chosen, the complex process of sketching each one along with its name began. The challenge was to illustrate each dish in an organic yet distinct and recognizable style, while keeping the illustration small enough to fit within its place on the map.

The idea on an aesthetic level was to create a map that feels not only like a gastronomic infographic but also like a hand-illustrated travel log. This is why each dish is accompanied by its name in handwritten letters. The various notes are written as if they were those of a traveler discovering the flavors and anecdotes of the Argentine land.

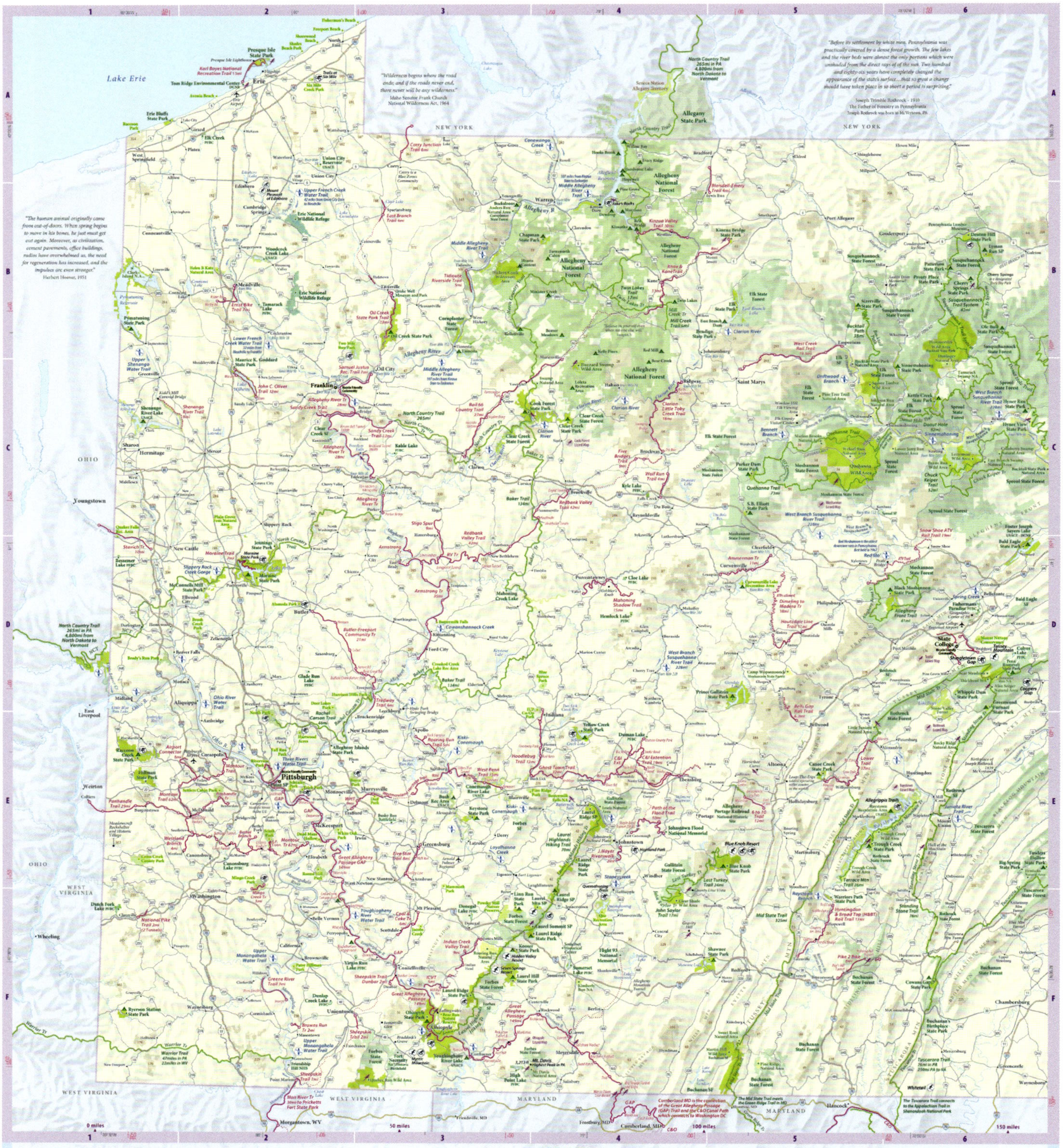

Purple Lizard Maps and Pennsylvania

At Purple Lizard, our mission is simple: to provide you with the best quality maps to guide and inspire your adventures. As a company founded in PA, we have been creating PA trail maps since 1997. We designed this map to be your statewide planning tool that opens up the Commonwealth of public lands and recreational resources in Pennsylvania and connects you with beautiful landscapes and the communities that surround them. We hope it will inspire many new outdoor adventures. Like you, we are committed to doing our part to keep these lands open, clean, and protected for future generations. Thank you for doing your part, and for making Purple Lizard Maps a part of your outdoor adventures. We'll see you out there!

"To me a lush carpet of pine needles or spongy grass is more welcome than the most luxurious Persian rug."
Helen Keller

This map is an invitation

We welcome you to explore all that Pennsylvania has to offer for outdoor recreation—hiking, biking (road, trail, gravel, mountain), paddling (flat or whitewater: canoe, kayak, paddleboard, or packraft), bird watching, camping, chilling, elk viewing, sailing, urban exploring, geocaching—whatever floats your proverbial boat.

Use this map as inspiration for what you want to do and where you want to do it. You won't be disappointed.

Get Involved

As you enjoy adventures in PA, we at the Pennsylvania Environmental Council (PEC) and Purple Lizard Maps want to encourage you to give a little back to the resource by making a commitment to the lands and waters that you enjoy. We ask that you take an active role as a steward of these resources.

To help you do that, we respectfully ask that you consider the following:

1. Join a local trail/watershed/conservation/recreation group and volunteer or attend an event.
2. Introduce someone to *their* local trail.
3. Visit a trail that's new to you.
4. Participate in at least one stewardship project a year.
5. Contact your local legislators about relevant legislation. Check out PEC's Bill Tracker at pecpa.org to stay up to date.

We also ask that you join us in recognizing and appreciating all the work that our public land managers and agencies do every day, to respect that different lands have different purposes, and to understand that your favorite recreational pursuit may not be available everywhere. The Pennsylvania Department of Conservation and Natural Resources, the Pennsylvania Game Commission, the Pennsylvania Fish and Boat Commission, the U.S. Forest Service, the National Park Service, county and local park agencies, and not-for-profit land holders are all front-line stewards of our public trust.

pec
pennsylvania environmental council

The people have a right to clean air, pure water, and the preservation of the natural, scenic, historic and esthetic values of the environment. Pennsylvania's public natural resources are the common property of all the people, including generations yet to come.
Article 1, Section 27, Pennsylvania Constitution

Pennsylvania State Parks

Allegheny Islands SP	E2
Bald Eagle SP	D6
Bendigo SP	B5
Big Spring SP	E6
Black Moshannon SP	D6
Blue Knob SP	E5
Buchanan's Birthplace SP	F6
Bucktail SP and Natural Area	C5,6
Canoe Creek SP	E5
Chapman SP	B4
Cherry Springs SP	B6
Clear Creek SP	C4
Cook Forest SP	C3
Cowans Gap SP	F6
Denton Hill SP	B6
Elk SP	B5
Erie Bluffs SP	A1
Fowlers Hollow SP	E6
Greenwood Furnace SP	D6
Hillman SP	E1
Hyner Run SP	C6
Hyner View SP	C6
Jennings SP	D2
Kettle Creek SP	C6
Keystone SP	E3
Kinzua Bridge SP	B5
Kooser SP	F3
Laurel Hill SP	F3
Laurel Mountain SP	F4
Laurel Ridge SP	E4; F3,4
Laurel Summit SP	F4
Linn Run SP	F3
Lyman Run SP	B6
Maurice K. Goddard SP	C2
McConnells Mill SP	D2
Moraine SP	D2
Ohiopyle SP	F3
Oil Creek SP	B3
Ole Bull SP	B6
Parker Dam SP	C5
Patterson SP	B6
Penn Roosevelt SP	D6
Point SP	E2
Presque Isle SP	A2
Prince Gallitzin SP	D5
Prouty Place SP	B6
Pymatuning SP	B1
Raccoon Creek SP	E1
Ryerson Station SP	F1
Shawnee SP	F4,5
Simon B. Elliott SP	C5
Sinnemahoning SP	C6
Sizerville SP	B5
Trough Creek SP	E5
Warriors Path SP	F5
Whipple Dam SP	D6
Yellow Creek SP	E4

PENNSYLVANIA Parks & Forests FOUNDATION

Inspiring stewardship of Pennsylvania's state parks and forests while working to ensure a place and an experience for everyone in the outdoors.

Long Distance Hiking Trails

Allegheny Front Trail	40.6 miles	D5,6
Baker Trail	133.5	C3,4; D3,4
Bucktail Path	38	C5
Chuck Keiper Trail	51.8	C6
Donut Hole Trail	81.7	C6
John Saylor Trail	17	E4
Laurel Highlands Hiking Trail	70.1	E4; F3,4
Lost Turkey Trail	23.5	E4,5
Mid State Trail	324.2	D6; E5,6; F5
Mill Creek & Twin Lakes Tr	5.6, 17.4	B4
North Country Trail	265 in PA	A4; B3,4; C2-4; D1,2
Quehanna Trail	73	C5,6
Rachel Carson Trail	45.7	D3; E2,3
Standing Stone Trail	75.7	D-F6
Susquehannock Trail System	82.1	B6; C6
Terrace Mountain Trail	25.9	E5,6
Tuscarora Trail	76.4 in PA	E6; F6
Warrior Trail	47.1 in PA	F1,2

State and National Forests

Allegheny National Forest	A4; B3,4; C4
Bald Eagle State Forest	D6
Buchanan State Forest	F4-6
Clear Creek State Forest	C2-4
Cornplanter State Forest	B3
Elk State Forest	B5; C5,6
Forbes State Forest	E3; F3,4
Gallitzin State Forest	E4; F4
Moshannon State Forest	C5; D5,6
Rothrock State Forest	D6; E5,6; F6
Sproul State Forest	C6
Susquehannock State Forest	B5,6
Tuscarora State Forest	E6; F6

Rail Trails and Bikeways over 5 miles in length

Rail Trails and Bikeways	Mileage	Surface	
6 to 10 Trail (part of this is hiking only)	12 miles	*Crushed Stone, Dirt*	E5
Allegheny River Trail	27.5	*Asphalt*	C2,3
Armstrong Trail	35.5	*Asphalt, Crushed Stone*	D3
Bells Gap Rail Trail	6.3	*Crushed Stone, Dirt, Grass*	D5; E5
Butler-Freeport Community Trail	21	*Asphalt, Crushed Stone, Gravel, Dirt*	D2,3
Clarion-Little Toby Creek Trail	18	*Crushed Stone*	C4
Coal & Coke Trail	6.1	*Asphalt, Crushed Stone*	F3
Corry Junction Greenway Trail	5.8	*Crushed Stone*	A3
David S. Ammerman Trail	10.5	*Crushed Stone*	D5
Dimeling to Madera Trail	18.3	*Unimproved*	D5
Ernst Bike Trail	7.2	*Asphalt, Gravel*	B2
Five Bridges Trail	8.5	*Ballast, Cinder, Dirt, Grass*	C4
Five Star Trail	7.8	*Crushed Stone*	E3
Ghost Town Trail	32	*Crushed Stone*	E4
Ghost Town: C&I Extension	19 when done	*Crushed Stone, Gravel, Dirt*	E4
Great Allegheny Passage (GAP)	148.8	*Asphalt, Crushed Stone, Gravel, Dirt*	E2; F2-4
Greene River Trail	7.2	*Crushed Stone*	F2
Hoodlebug Trail	11.8	*Asphalt, Gravel*	D4; E3,4
Houtzdale Line Trail	11	*Gravel, Dirt*	D5
Huntingdon & Broad Top RT (H&BT)	12.8	*Gravel, Dirt*	F5
Indian Creek Valley Trail	19 when done	*Crushed Stone*	F3
John C. Oliver Trail	12 miles	*Asphalt*	C2
Karl Boyes National Recreation Trail	13.4	*Asphalt*	A2
Kinzua Valley Trail	10	*Crushed Stone, Gravel, Dirt*	B4
Knox & Kane Rail Trail	13 & 7.8	*Asphalt, Crushed Stone*	B4,5
Lower Trail (pronounced like "flower")	16.5	*Crushed Stone, Asphalt*	E5
Mahoning Shadow Trail	15	*Asphalt, Crushed Stone*	D4
Montour Trail	61.5	*Asphalt, Crushed Stone*	E2
Montour Trail - Airport Connector	6.4	*Crushed Stone*	E2
Oil Creek State Park Trail	12.8	*Asphalt*	B3; C3
Panhandle Trail	29.2	*Asphalt, Crushed Stone*	E1,2
Path of the Flood Trail	10	*Asphalt, Crushed Stone, Gravel, Dirt*	E4
Pike 2 Bike	8.2	*Asphalt*	F5
Rail 66 Country Trail	27	*Asphalt*	C3,4
Redbank Valley Trail	42	*Crushed Stone*	C3,4; D3
Redbank Valley Trail - Sligo Spur	9	*Gravel, Dirt*	C3; D3
Samuel Justus Rec Trail	7	*Asphalt*	C2,3
Sandy Creek Trail	12	*Asphalt*	C2,3
Shenango River Trail	9	*Gravel, Dirt*	C1
Snow Shoe Rail Trail	18.5	*Gravel, Dirt*	D5,6
Stavich Trail	10	*Asphalt*	C1; D1
West Creek Rail Trail	19.5	*Crushed Stone*	B5; C5
West Penn Trail	13	*Asphalt, Crushed Stone, Gravel, Dirt*	E3
Westmoreland Heritage Trail	9.3 & 8.5	*Crushed Stone*	E2,3

Acquiring parks and open spaces...in and around our urban centers must be regarded as a 'now or never' proposition...[it] simply cannot be postponed until sometime in the remote and hazy future, because land costs in these areas, high as they are now, will be too high for government purchase in another 10 years.

M. K. Goddard
Secretary of the Pennsylvania
Department of Forests and Waters
1955-1970

"Let the clean air blow the cobwebs from your body. Air is medicine."
Lillian Russell (1862-1922)
USDA Forest Service

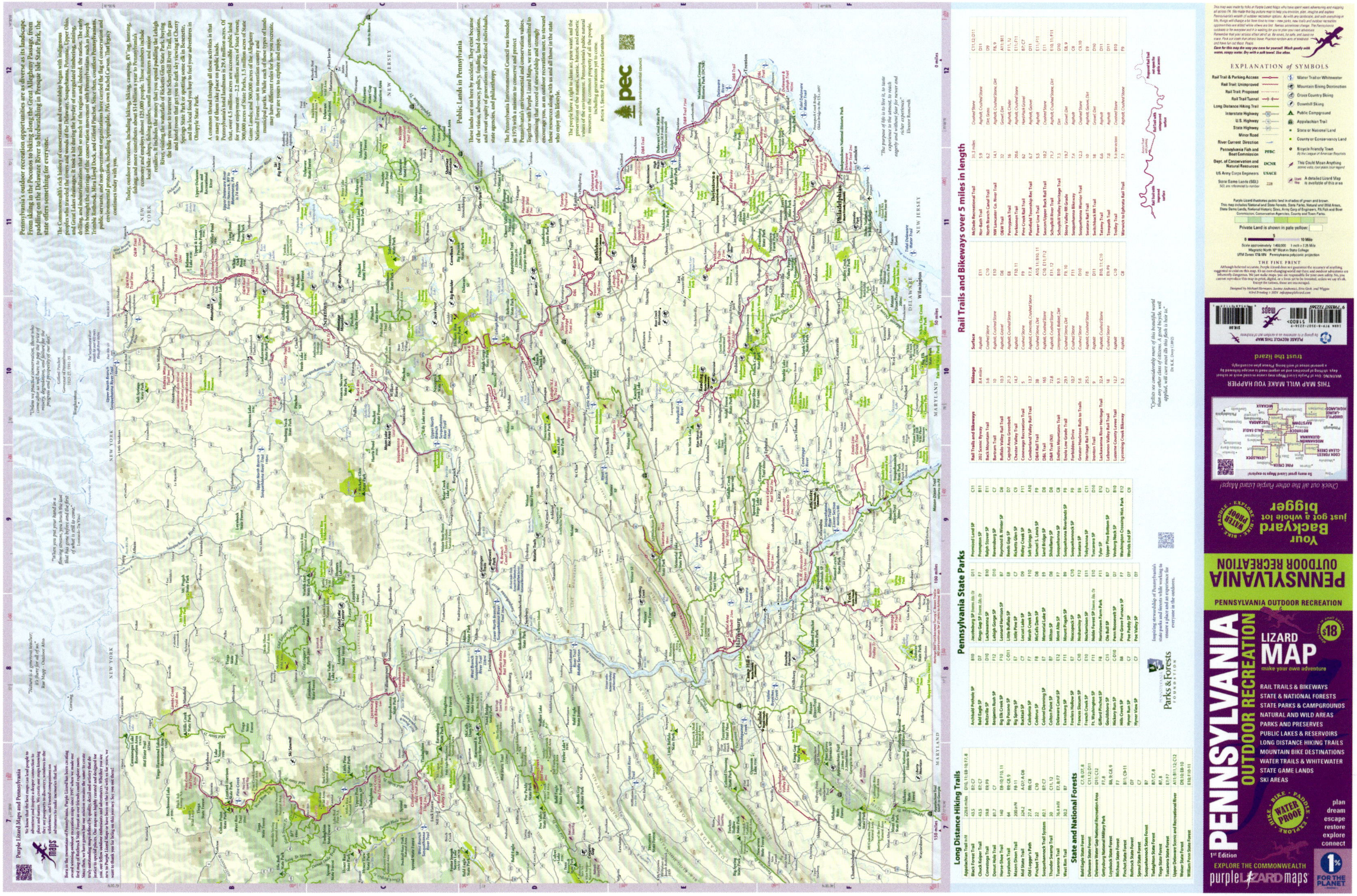
PENNSYLVANIA
OUTDOOR RECREATION
LIZARD MAP
make your own adventure
$18
RAIL TRAILS & BIKEWAYS
STATE & NATIONAL FORESTS
STATE PARKS & CAMPGROUNDS
NATURAL AND WILD AREAS
PARKS AND PRESERVES
PUBLIC LAKES & RESERVOIRS
LONG DISTANCE HIKING TRAILS
MOUNTAIN BIKE DESTINATIONS
WATER TRAILS & WHITEWATER
STATE GAME LANDS
SKI AREAS
plan
dream
escape
restore
explore
connect
1st Edition
EXPLORE THE COMMONWEALTH
purple lizard maps
1% FOR THE PLANET
PENNSYLVANIA OUTDOOR RECREATION
Your Backyard just got a whole lot bigger
THIS MAP WILL MAKE YOU HAPPIER
EXPLANATION of SYMBOLS
Rail Trails and Bikeways over 5 miles in length
Pennsylvania State Parks
Long Distance Hiking Trails
State and National Forests
Public Lands in Pennsylvania
Parks & Forests FOUNDATION
pec
NEW YORK
NEW JERSEY
MARYLAND
DELAWARE
Philadelphia
Harrisburg
Scranton
Reading
Wilmington
Trenton

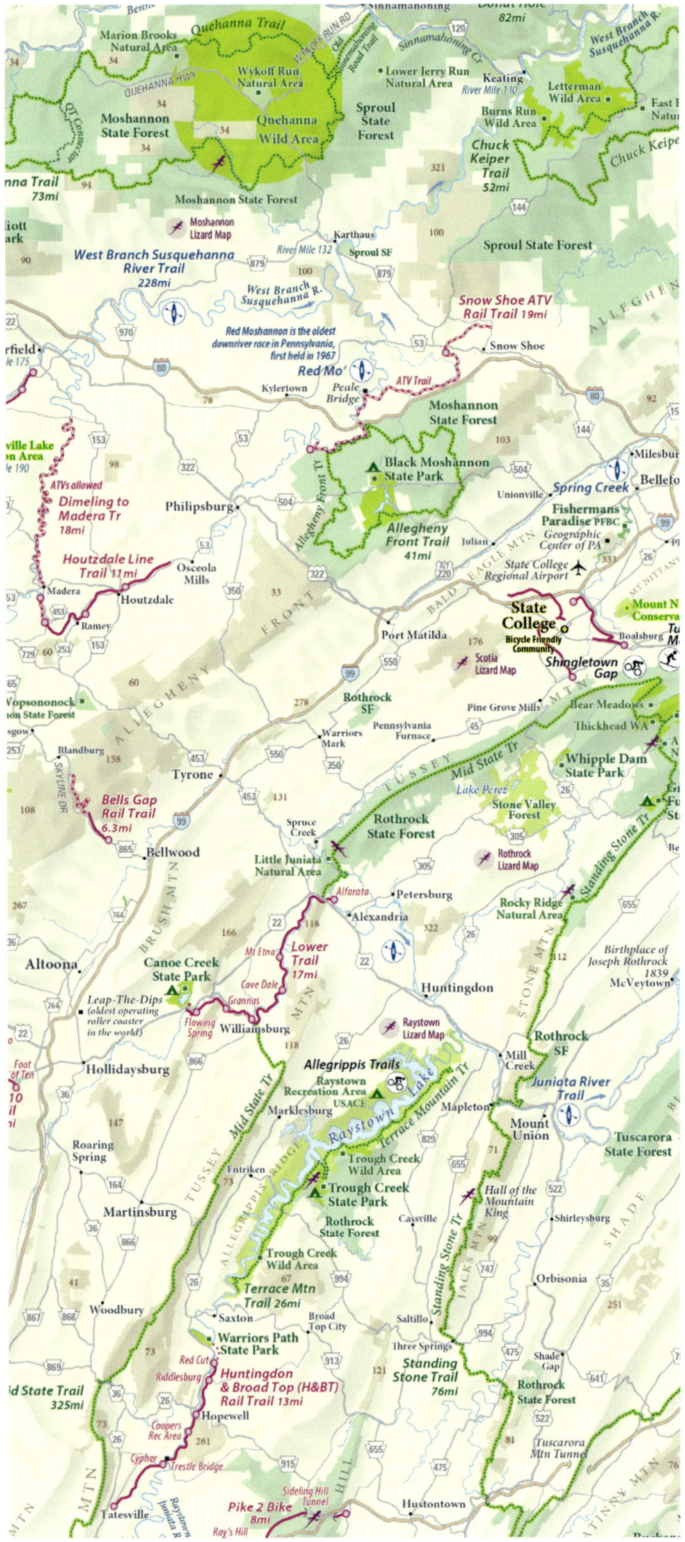

Quehanna Trail
Marion Brooks Natural Area
Wykoff Run Natural Area
Moshannon State Forest
Quehanna Wild Area
Sproul State Forest
Lower Jerry Run Natural Area
Burns Run Wild Area
Chuck Keiper Trail 52mi
Letterman Wild Area
Moshannon State Forest
Moshannon Lizard Map
Karthaus
Sproul State Forest
West Branch Susquehanna River Trail 228mi
Snow Shoe ATV Rail Trail 19mi
Snow Shoe
Red Moshannon is the oldest downriver race in Pennsylvania, first held in 1967
Red Mo'
Kylertown
Peale Bridge
ATV Trail
Moshannon State Forest
Black Moshannon State Park
Spring Creek
Allegheny Front Trail 41mi
Philipsburg
Unionville
Fishermans Paradise PFBC
Geographic Center of PA
Julian
State College Regional Airport
Dimeling to Madera Tr 18mi
ATVs allowed
Houtzdale Line Trail 11mi
Osceola Mills
Houtzdale
Madera
Ramey
Port Matilda
State College
Bicycle Friendly Community
Boalsburg
Scotia Lizard Map
Shingletown Gap
Mount Nittany Conservancy
Rothrock SF
Warriors Mark
Pennsylvania Furnace
Pine Grove Mills
Bear Meadows
Thickhead WA
Whipple Dam State Park
Mid State Tr
Lake Perez
Stone Valley Forest
Tyrone
Blandburg
Bells Gap Rail Trail 6.3mi
Rothrock State Forest
Rothrock Lizard Map
Spruce Creek
Little Juniata Natural Area
Bellwood
Rocky Ridge Natural Area
Standing Stone Tr
Alfarata
Petersburg
Alexandria
Birthplace of Joseph Rothrock 1839
McVeytown
Altoona
Canoe Creek State Park
Lower Trail 17mi
Mt Etna
Cove Dale
Grannas
Huntingdon
Leap-The-Dips (oldest operating roller coaster in the world)
Flowing Spring
Williamsburg
Raystown Lizard Map
Rothrock SF
Mill Creek
Hollidaysburg
Allegrippis Trails
Raystown Recreation Area USACE
Juniata River Trail
Marklesburg
Mapleton
Mount Union
Raystown Lake
Terrace Mountain Tr
Tuscarora State Forest
Roaring Spring
Entriken
Trough Creek Wild Area
Trough Creek State Park
Rothrock State Forest
Hall of the Mountain King
Martinsburg
Cassville
Shirleysburg
Trough Creek Wild Area
Terrace Mtn Trail 26mi
Orbisonia
Woodbury
Saxton
Broad Top City
Warriors Path State Park
Saltillo
Three Springs
Standing Stone Trail 76mi
Shade Gap
Rothrock State Forest
Red Cut
Riddlesburg
Huntingdon & Broad Top (H&BT) Rail Trail 13mi
Mid State Trail 325mi
Hopewell
Coopers Rec Area
Cypher
Trestle Bridge
Tuscarora Mtn Tunnel
Sideling Hill Tunnel
Pike 2 Bike 8mi
Ray's Hill
Tatesville
Hustontown

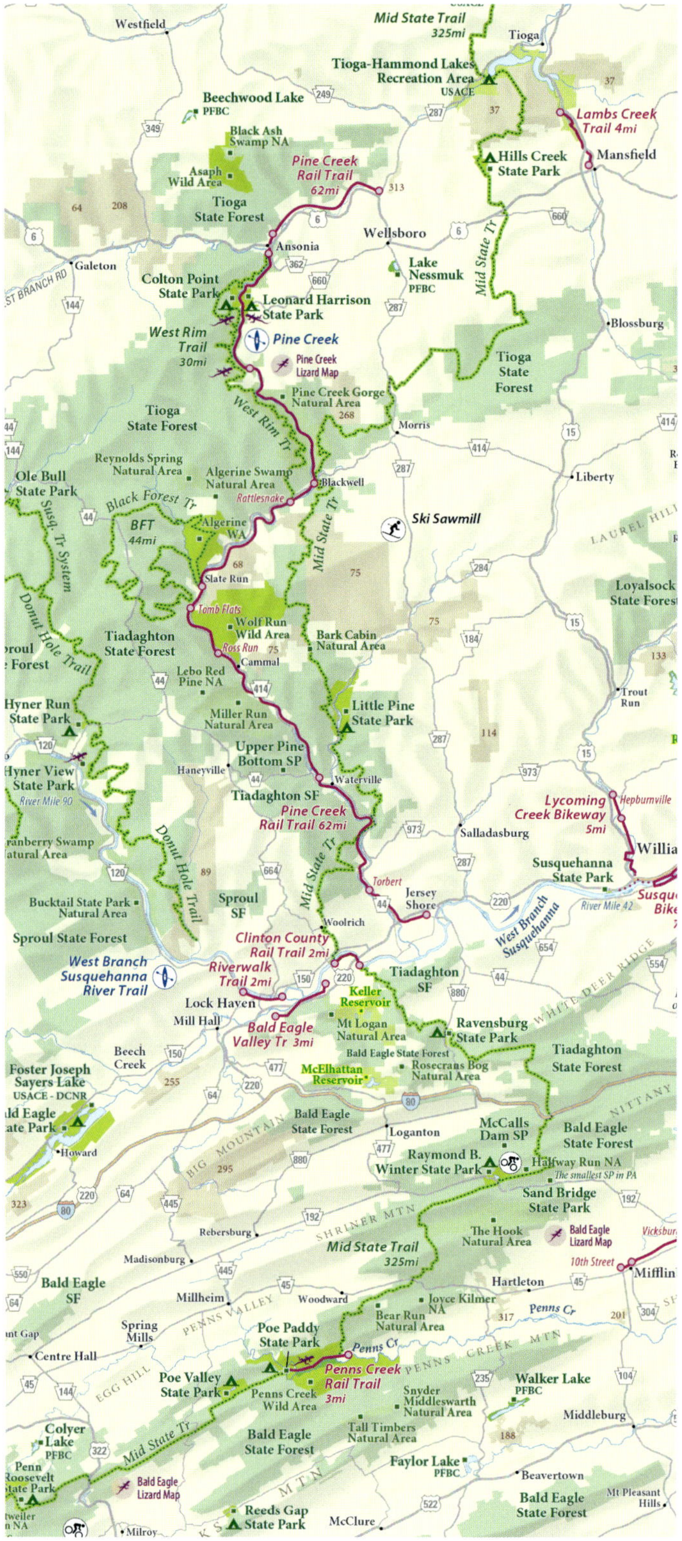

Mid State Trail 325mi
Westfield
Tioga
Tioga-Hammond Lakes Recreation Area USACE
Beechwood Lake PFBC
Lambs Creek Trail 4mi
Black Ash Swamp NA
Pine Creek Rail Trail 62mi
Hills Creek State Park
Mansfield
Asaph Wild Area
Tioga State Forest
Wellsboro
Galeton
Ansonia
Colton Point State Park
Leonard Harrison State Park
Lake Nessmuk PFBC
Mid State Tr
West Rim Trail 30mi
Pine Creek
Pine Creek Lizard Map
Blossburg
Tioga State Forest
Pine Creek Gorge Natural Area
Tioga State Forest
Morris
Reynolds Spring Natural Area
Algerine Swamp Natural Area
Blackwell
Liberty
Ole Bull State Park
Black Forest Tr
Rattlesnake
Ski Sawmill
BFT 44mi
Algerine WA
Slate Run
Tomb Flats
Wolf Run Wild Area
Ross Run
Tiadaghton State Forest
Bark Cabin Natural Area
Loyalsock State Forest
Donut Hole Trail
Cammal
Lebo Red Pine NA
Hyner Run State Park
Miller Run Natural Area
Little Pine State Park
Upper Pine Bottom SP
Haneyville
Waterville
Hyner View State Park
Tiadaghton SF
Pine Creek Rail Trail 62mi
Lycoming Creek Bikeway 5mi
Hepburnville
Cranberry Swamp Natural Area
Salladasburg
Susquehanna State Park
Torbert
Jersey Shore
Bucktail State Park Natural Area
Sproul SF
Woolrich
Sproul State Forest
Clinton County Rail Trail 2mi
West Branch Susquehanna River Trail
Riverwalk Trail 2mi
Lock Haven
Mill Hall
Tiadaghton SF
Keller Reservoir
Bald Eagle Valley Tr 3mi
Mt Logan Natural Area
Ravensburg State Park
Bald Eagle State Forest
Rosecrans Bog Natural Area
Tiadaghton State Forest
Beech Creek
McElhattan Reservoir
Foster Joseph Sayers Lake USACE - DCNR
Bald Eagle State Park
Howard
Bald Eagle State Forest
Loganton
McCalls Dam SP
Bald Eagle State Forest
Raymond B. Winter State Park
Halfway Run NA
The smallest SP in PA
Sand Bridge State Park
Rebersburg
The Hook Natural Area
Bald Eagle Lizard Map
Mid State Trail 325mi
Vicksburg
Mifflin
Madisonburg
Hartleton
Bald Eagle SF
Millheim
Woodward
Joyce Kilmer NA
Bear Run Natural Area
Spring Mills
Poe Paddy State Park
Penns Cr
Centre Hall
Poe Valley State Park
Penns Creek Rail Trail 3mi
Walker Lake PFBC
Penns Creek Wild Area
Snyder Middleswarth Natural Area
Colyer Lake PFBC
Bald Eagle State Forest
Tall Timbers Natural Area
Middleburg
Penn Roosevelt State Park
Faylor Lake PFBC
Beavertown
Bald Eagle Lizard Map
Reeds Gap State Park
Bald Eagle State Forest
McClure
Milroy

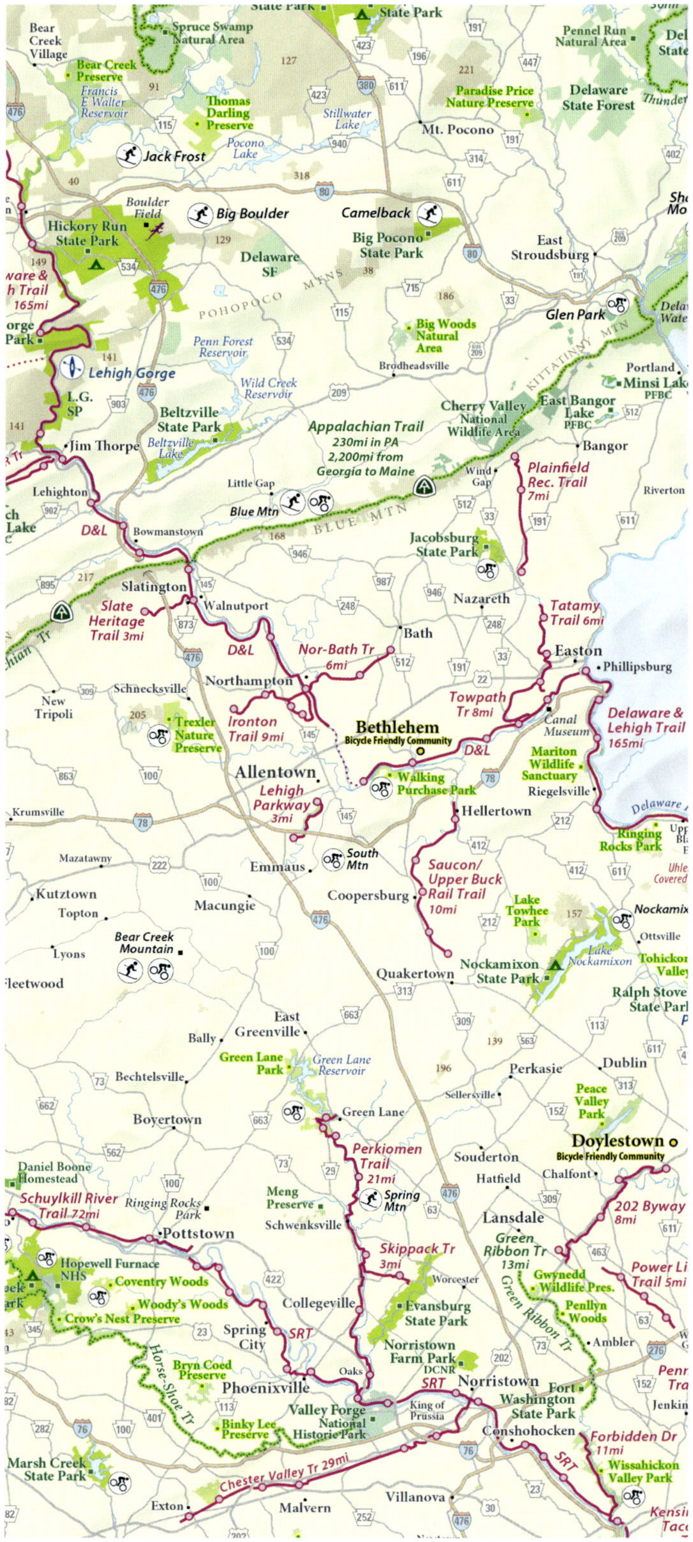

State Park
Bear Creek Village
Spruce Swamp Natural Area
Pennel Run Natural Area
Bear Creek Preserve
Francis E. Walter Reservoir
Paradise Price Nature Preserve
Delaware State Forest
Thomas Darling Preserve
Stillwater Lake
Mt. Pocono
Jack Frost
Pocono Lake
Boulder Field
Big Boulder
Camelback
Hickory Run State Park
Delaware SF
Big Pocono State Park
East Stroudsburg
Lehigh Gorge
Big Woods Natural Area
Glen Park
Penn Forest Reservoir
Brodheadsville
Portland
Minsi Lake PFBC
L.G. SP
Wild Creek Reservoir
Beltzville State Park
Beltzville Lake
Appalachian Trail 230mi in PA 2,200mi from Georgia to Maine
Cherry Valley National Wildlife Area
East Bangor Lake PFBC
Jim Thorpe
Bangor
Plainfield Rec. Trail 7mi
Lehighton
Little Gap
Blue Mtn
Wind Gap
D&L
Bowmanstown
Jacobsburg State Park
Slatington
Walnutport
Nazareth
Tatamy Trail 6mi
Slate Heritage Trail 3mi
Bath
Nor-Bath Tr 6mi
Easton
Phillipsburg
Northampton
Schnecksville
New Tripoli
Trexler Nature Preserve
Ironton Trail 9mi
Bethlehem
Bicycle Friendly Community
Towpath Tr 8mi
Canal Museum
Delaware & Lehigh Trail 165mi
Allentown
Walking Purchase Park
Marlton Wildlife Sanctuary
Riegelsville
Lehigh Parkway 3mi
Hellertown
Krumsville
Emmaus
South Mtn
Ringing Rocks Park
Saucon/Upper Buck Rail Trail 10mi
Kutztown
Macungie
Coopersburg
Lake Towhee Park
Nockamixon
Topton
Lyons
Bear Creek Mountain
Quakertown
Nockamixon State Park
Ralph Stover State Park
Fleetwood
East Greenville
Bally
Green Lane Park
Green Lane Reservoir
Perkasie
Dublin
Bechtelsville
Peace Valley Park
Boyertown
Green Lane
Doylestown
Bicycle Friendly Community
Perkiomen Trail 21mi
Souderton
Daniel Boone Homestead
Spring Mtn
Hatfield
Chalfont
Schuylkill River Trail 72mi
Ringing Rocks Park
Meng Preserve
Schwenksville
Lansdale
202 Byway 8mi
Pottstown
Green Ribbon Tr 13mi
Hopewell Furnace NHS
Coventry Woods
Skippack Tr 3mi
Gwynedd Wildlife Pres.
Power Line Trail 5mi
Woody's Woods
Collegeville
Evansburg State Park
Penllyn Woods
Crow's Nest Preserve
Spring City
SRT
Ambler
Norristown Farm Park DCNR
Bryn Coed Preserve
Horse-Shoe Tr
Phoenixville
Norristown
Fort Washington State Park
Valley Forge National Historic Park
Binky Lee Preserve
King of Prussia
Conshohocken
Forbidden Dr 11mi
Marsh Creek State Park
Wissahickon Valley Park
Chester Valley Tr 29mi
Exton
Malvern
Villanova

« Purple Lizard Pennsylvania Outdoor Recreation Map

Michael Hermann, Justine Andronici, Erin Greb, and Bill Wilgus

The best recreation maps don't just guide people; they inspire a deeper connection with place and spark the imagination for future adventures. This map is both a tribute to an extraordinary outdoor landscape and a vision of what's possible. In a time of fixation on 2-inch-wide screens that limit our ability to see the full landscape, this map connects the dots across the abundance of outdoor recreation options in the Commonwealth of Pennsylvania.

We largely used a combination of color and typography to identify and separate the wealth of Pennsylvania outdoor recreation assets. These include hundreds of state parks, rail trails, bikeways, mountain bike destinations, long distance hiking trails, ski areas, water trails, and whitewater destinations around and within state and national forests, recreation areas, wild and natural areas, state game lands, and public lakes managed by the Pennsylvania Fish and Boat Commission and the Army Corps of Engineers.

Determining the visual hierarchies of so many diverse recreational assets pushed us toward developing an increasingly subtle color palette. With a goal of softening the map aesthetic, we pushed items such as highway badges and roads further in the background. The map's clarity is also enhanced by what the cartographers chose not to include: township and county lines, secondary roads, and smaller tributaries, for instance. The extensive network of bikeways and rail trails was given the strongest visual weight, with long-distance hiking trails a close second.

This map was designed for the retail market to benefit people interested in discovering new places to pursue outdoor recreation. It also serves the needs of planners, politicians, and environmentalists looking for opportunities to connect people and communities with bike paths, hiking trails, and public lands, and helps people visualize the potential for Pennsylvania's impressive future as a leading state in outdoor recreation. In that way, the true power of this map is how it can work to improve the quality of life for both residents of and visitors to the Commonwealth of Pennsylvania.

Spirits of Scotland »

Milana Glebova

The idea for this map emerged while I was reading *Raw Spirit: In Search of the Perfect Dram* by Iain Banks, a book recounting his visits to Scottish whisky distilleries. To follow the author's journey across lochs, fjords, and sounds, I often paused to search for the places he described. This became increasingly frustrating as I wanted to quickly navigate the distilleries and the winding routes between them, which is not particularly easy to do on digital maps where the distilleries are only visible at detailed zoom levels. The narrative in the book was also tangible enough that I wished to browse a detailed map while reading and chart a route with a pencil (for future travels, of course).

The first spatial puzzle to solve was a wish to have all the locations labeled, yet to avoid cluttered icons, while keeping the map a reasonable size to hang on a wall. Other maps have tried to mitigate this problem by using numbers keyed to a long list of names, but I didn't like how this layout distracts the reader from the map itself and makes them spend time looking through a list. During the process, I had the notion to transform humble leading lines into decorative features, which ultimately took center stage in the map's aesthetic. Apart from creating a pattern that resembles the outlines of Glencairn whisky glasses, they color-code whisky regions: Highland, Speyside, Lowland, Campbeltown, Islay, and Islands.

Visually, I meant to keep the relief ductility and imitate a slightly old-fashioned, velvety feel of mountain ridges and valleys so they were warmly highlighted and so the image doesn't look digitally produced. The terrain is a layered pie where a land polygon is overlaid with shaded relief layers colored in a gold-to-blue gradient, and a semitransparent mist gives valleys and lowlands depth.

Researching the finer details of a map and inevitably getting lost in the stories associated with them is, for me, the closest thing to traveling, especially in the case of places of longing. This pet project was intertwined with that idea: it started as a tool for self-tracking a book journey and turned into a way to create something beautiful—a declaration of love for Scotland.

spirits of scotland

This map showcases the amazing whisky distilleries sprinkled across Scotland. The colors help distinguish the regions of whisky production: **Highland**, **Speyside**, **Lowland**, **Campbeltown**, **Islay**, and **Islands** (which is officially a part of the Highland region, but is sometimes distinguished as a region of its own).

So, whether you're plotting your next whisky trail or just satisfying Scottish wanderlust, here's your guide. Remember, it's all about the journey (and the drams).

Slainte!

No village left unscathed: In brutal attack, Hamas terrorists kill more than 1000 people in areas surrounding the Gaza Strip

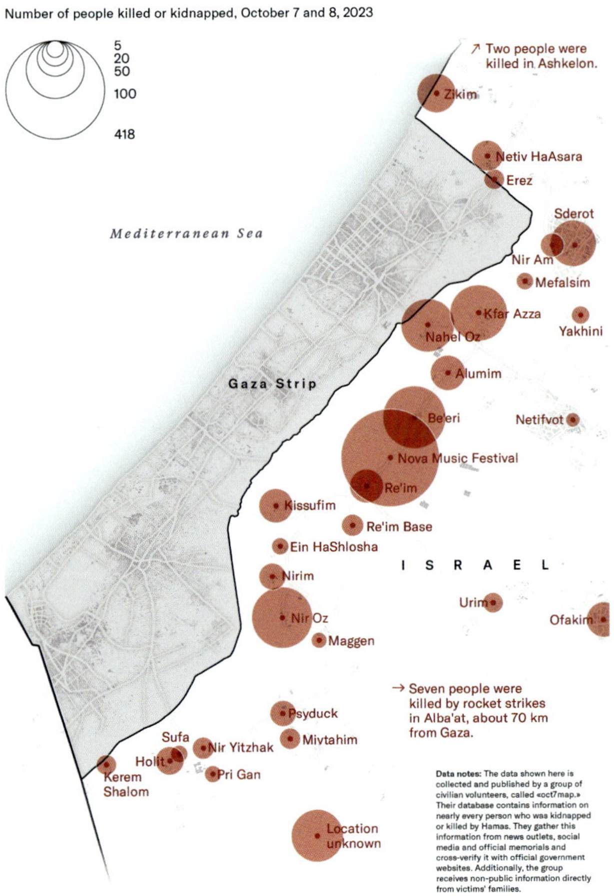

Open-air prison before the war: An Israeli land and sea blockade envelops the Gaza Strip

Diminishing havens: Safe spaces for civilians shrink amid Israeli ground invasion

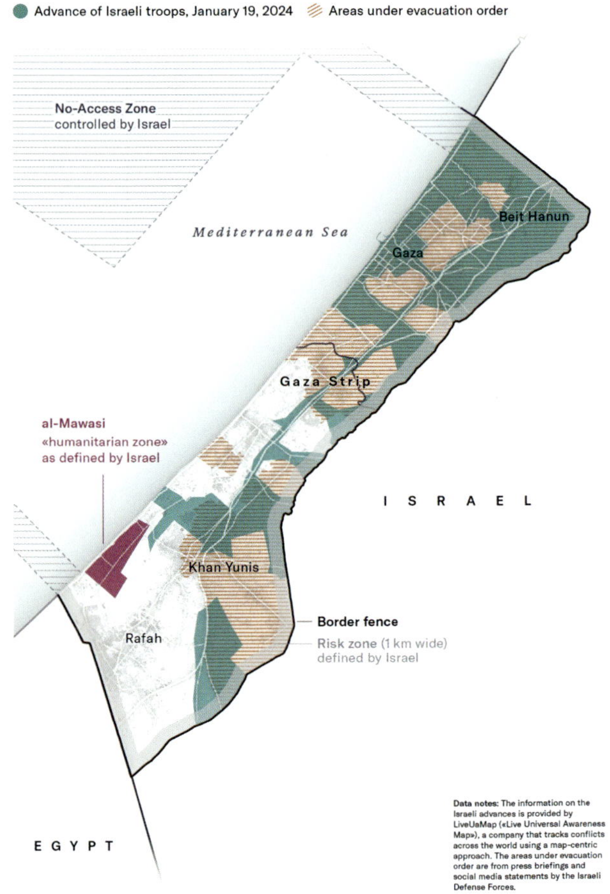

Under Siege: A Cartographic Chronicle of the Fifth Gaza War »

Adina Renner, Neue Zürcher Zeitung

In recent years, detailed online maps have become central to conflict reporting at the Swiss daily *Neue Zürcher Zeitung*. The Gaza war is no exception. Unlike the war in Ukraine, where precise troop movements capture readers' interest, the Israel-Palestine conflict calls for diverse thematic maps that illuminate various facets of an intractable situation. This series presents five such maps.

In their sequence, the maps reflect the progression of the fifth Gaza war. We begin by focusing on the areas outside of the Gaza Strip, identifying the locations and counts of those killed or kidnapped in the Hamas attack in October 2023. Next, we examine the Gaza Strip itself, seeking answers in its geographic and political isolation. Then, in November, the ground offensive starts. Israeli troops enter Gaza, ordering Palestinians to evacuate to a small strip of arid land. As the war continues, the destruction from shelling grows. Entire neighborhoods are leveled, and a humanitarian crisis unfolds: hospitals cease to function, and a diesel import ban disrupts electricity, communications, and water supplies.

Destruction from the skies: Roughly 50% of all buildings in the Gaza Strip have been damaged or destroyed by shelling

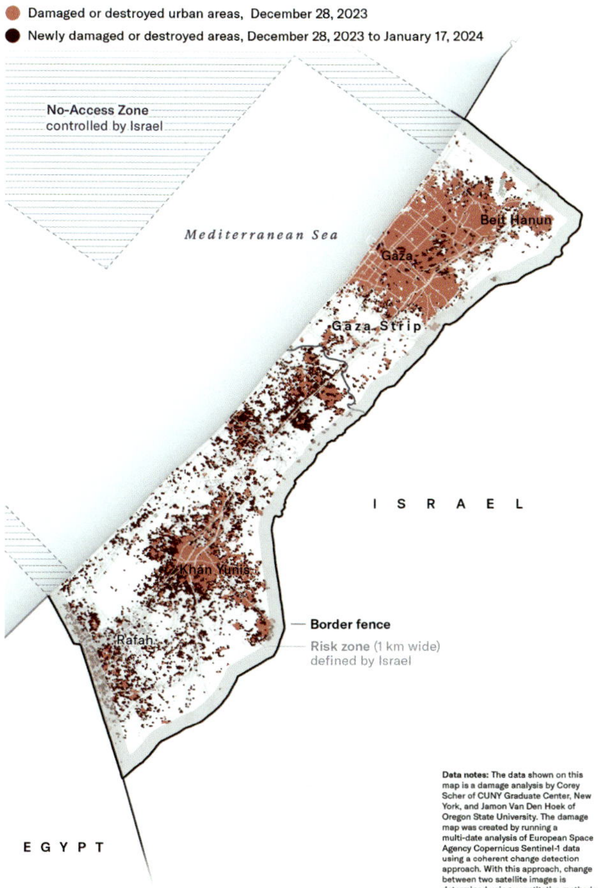

Data notes: The data shown on this map is a damage analysis by Corey Scher of CUNY Graduate Center, New York, and Jamon Van Den Hoek of Oregon State University. The damage map was created by running a multi-date analysis of European Space Agency Copernicus Sentinel-1 data using a coherent change detection approach. With this approach, change between two satellite images is determined using quantitative methods.

A dire humanitarian situation: Only around half of the hospitals in the Gaza Strip are still functional

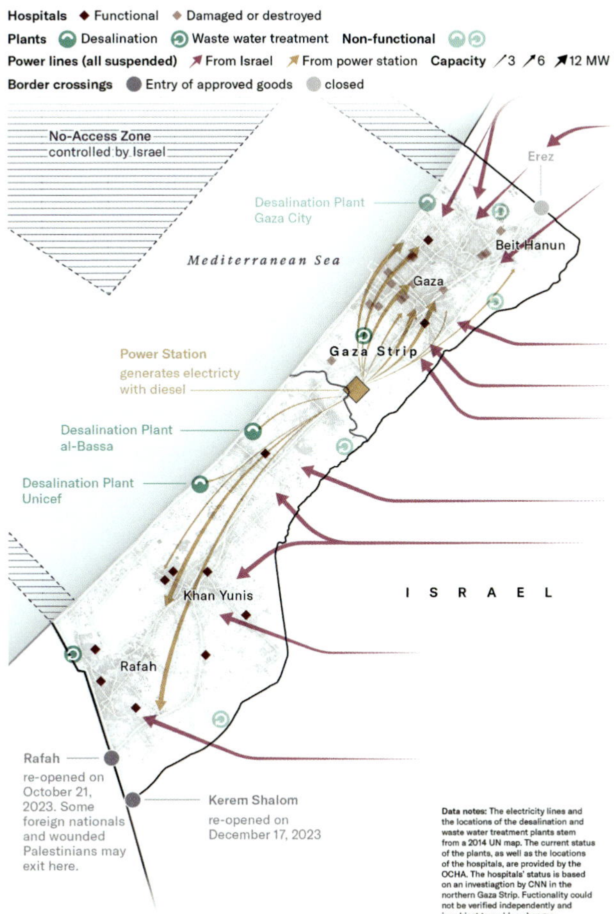

Data notes: The electricity lines and the locations of the desalination and waste water treatment plants stem from a 2014 UN map. The current status of the plants, as well as the locations of the hospitals, are provided by the OCHA. The hospitals' status is based on an investiagtion by CNN in the northern Gaza Strip. Fuctionality could not be verified independently and is subject to sudden changes.

Under siege: A cartographic chronicle of the fifth Gaza war

This series of maps was developed following the Hamas attack on October 7 and the subsequent war in Gaza. The collection of five maps offers a comprehensive view: They illustrate the scale of the attack, provide a geopolitical perspective of the Gaza Strip, examine its infrastructure, and track key developments of the conflict through January 2024.

These maps were produced for various online news articles published by the Swiss daily "Neue Zürcher Zeitung." Their purpose is to offer readers a concise and clear understanding of the complex and dynamic situation in the Middle East.

The Israel-Palestine conflict is a conflict fueled by maps. Lines drawn by colonial powers have destroyed livelihoods and broken promises. And yet, I have hope that maps can redeem their function as a helpful medium. In an era of information overload, maps have the capacity to convey complex geographic realities without emotional manipulation. Photographs capture human suffering but often remain anecdotal. Words alone cannot convey the full picture. If created with care, maps offer the potential to show the horrors of war while providing clear and educational information.

How can we create with care? Mapmaking involves numerous small decisions that culminate in a complete picture. For instance, in these maps, red is used to mark both those killed by Hamas and the destruction in Gaza caused by Israeli forces. The term "humanitarian zone" is placed inside quotation marks, reflecting Israeli designation. And the base map shows every house, reminding the reader of human presence. By attending to these layered decisions, maps can incorporate and convey varied perspectives on a conflict frequently depicted with bias and oversimplification.

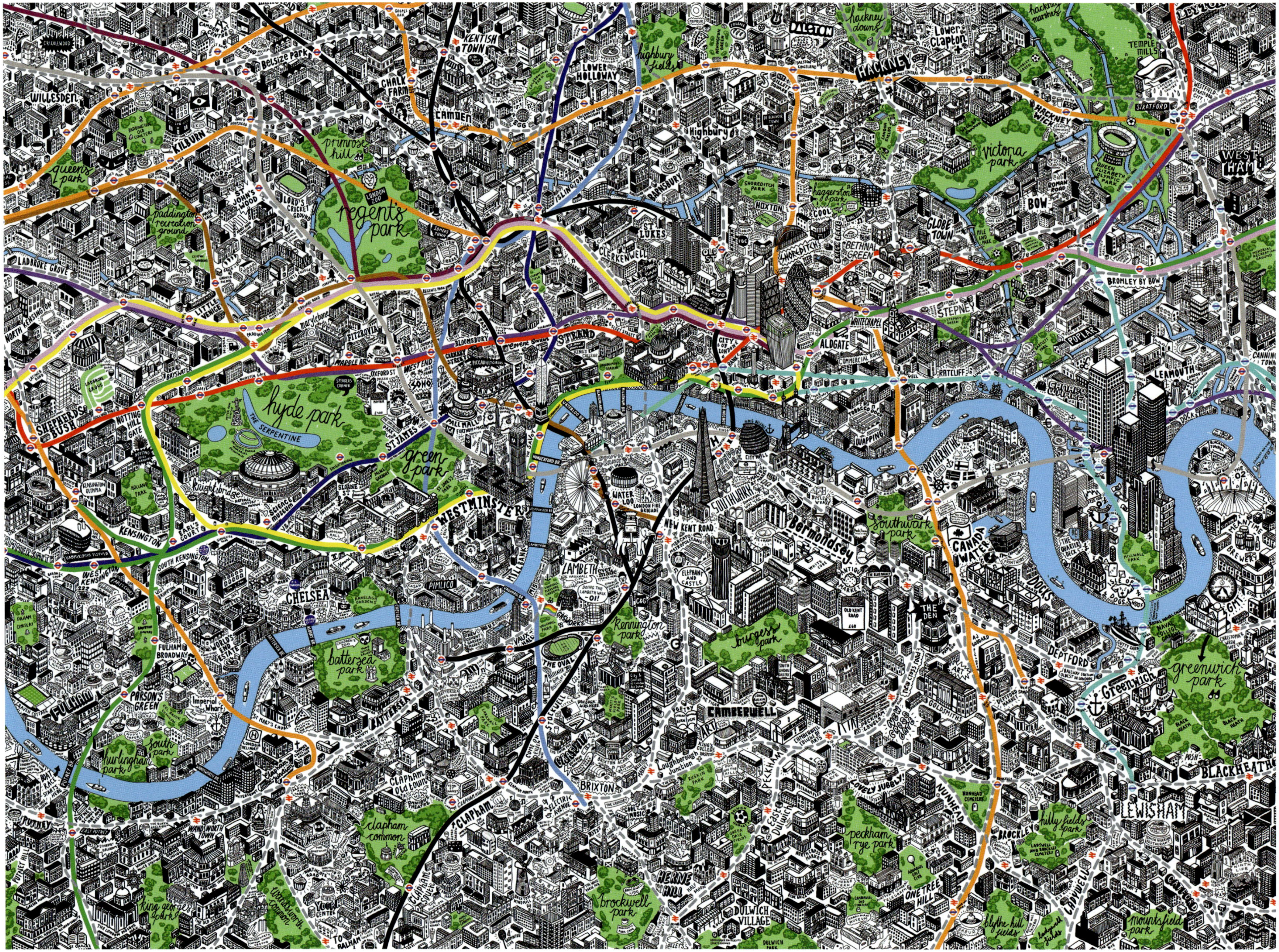

WILLESDEN
KILBURN
queen's park
Belsize Park
CHALK FARM
KENTISH TOWN
LOWER HOLLOWAY
highbury fields
highbury
DALSTON
hackney downs
HACKNEY
Lower Clapton
TEMPLE MILLS
STRATFORD
WEST HAM
CAMDEN
primrose hill
regent's park
paddington recreation ground
LADBROKE GROVE
victoria park
BOW
GLOBE TOWN
haggerston park
HOXTON
SHOREDITCH PARK
BETHNAL GREEN
ST. LUKES
CLERKENWELL
ALDGATE
WHITECHAPEL
STEPNEY
POPLAR
BROMLEY BY BOW
LEAMOUTH
SHEPHERD'S BUSH
hyde park
THE SERPENTINE
MARBLE ARCH
OXFORD ST
SOHO
ST JAMES
green park
WESTMINSTER
KENSINGTON
SOUTH KENSINGTON
CHELSEA
PIMLICO
LAMBETH
WATERLOO
NEW KENT ROAD
SOUTHWARK
Bermondsey
southwark park
CANADA WATER
WAPPING
ROTHERHITHE
SHADWELL
ELEPHANT AND CASTLE
kennington park
burgess park
OLD KENT ROAD
THE DEN
DEPTFORD
Greenwich
greenwich park
BLACKHEATH
battersea park
THE OVAL
CAMBERWELL
FULHAM
FULHAM BROADWAY
PARSON'S GREEN
south park
hurlingham park
BATTERSEA
BRIXTON
PECKHAM
peckham rye park
clapham common
CLAPHAM
CLAPHAM OLD TOWN
brockwell park
HERNE HILL
DULWICH VILLAGE
ONE TREE HILL
LEWISHAM
hilly fields park
BROCKLEY
mountsfield park
blythe hill fields
king george's park
wandsworth common
PUTNEY

« Hand Drawn Map of London

Jenni Sparks, Evermade

In 2012, when I was in my early twenties, I made the exciting (and overwhelming) move to London from a small town, hoping to establish a career in illustration. That same year, London hosted the Olympics and celebrated Queen Elizabeth's Diamond Jubilee, drawing the world's eyes to the city. It was during this time that Evermade, a fledgling art print company, approached me with an idea: to illustrate a map of London that transcended the trope of reducing a city to its tourist-friendly highlights, and could appeal to visitors and Londoners alike.

The idea of being paid to explore the city for research was appealing, since I had only recently settled there. There was also a part of me that, in making the map, wanted to feel accepted by London itself, aspiring to shed my small-town naïveté and evolve into a sophisticated urbanite.

And so I spent many memorable hours traversing the tube and walking through various neighborhoods, taking in their atmospheres and noting as many unique details as I could. I looked for things such as architectural landmarks, and felt what the energy was like—wealthy, artsy, rough around the edges. I also kept my eyes peeled for smaller details like a piece of street art or a funny shop sign.

I wanted the map to be funny at times, with a bit of typically British satire. I sought to infuse it with a sense of vibrancy, even a touch of irreverence, reflecting my own style and personality. With most of the map executed in black and white, the colorful tube and train lines provide a sense of structure to the chaos. I also used a hand-drawn font for each neighborhood that reflected the character of the place, as well as doodles—hearts, stars, and even stink lines—to jazz things up and create a sense of energy.

When the map was completed, it felt like the start of something big, and gave me a sense that I might be able to have a creative career after all. Nine years later, in 2021, Evermade commissioned me again to update various elements, including new landmarks and points of interest. By then, having lived in London for many years, I had grown up, and evolved alongside the city. My experiences there had multiplied, as had my knowledge.

East Boston Sea Level Rise Flood Risk »

Andy Woodruff

In 2023 I spent a month as designer-in-residence at the Norman B. Leventhal Map and Education Center (LMEC) at the Boston Public Library. This unique fellowship was created in support of the LMEC's goal of hosting not only historical maps and related resources, but also ongoing, modern cartography. The residency had only one requirement: make a map!

I was tasked with responding cartographically to the theme of the LMEC exhibition at the time, which highlighted the stories told by so-called urban atlases of the nineteenth and twentieth centuries, such as Sanborn fire insurance maps. These maps have lived beyond their original purposes, proving useful for historical urban research and offering fascinating windows into the life and form of cities at different times in history.

This map is the result of my desire to reproduce the look of those old maps using present-day information. Thanks to a suggestion by LMEC president and head curator Garrett Dash Nelson, it was more than just an exercise in style. Many of the old maps are maps of risk—that is, fire—and this, in turn, is a map of a modern risk: flooding from projected sea level rise scenarios in the 2016 Climate Ready Boston report.

While I had fun learning and working on various methods of digital fakery to make an antique-looking map, I also hope this map serves as something of a think piece. The map plates in old urban atlases are at neighborhood scales, and to my surprise are physically quite large. My map, a single plate in an imagined atlas, also shows only one section of one Boston neighborhood and also is designed for a large print size. Climate-change maps that most of us see are, at best, for a whole city or region. Perhaps a building-level map with property owner names like this could make climate-related risks seem more real and present.

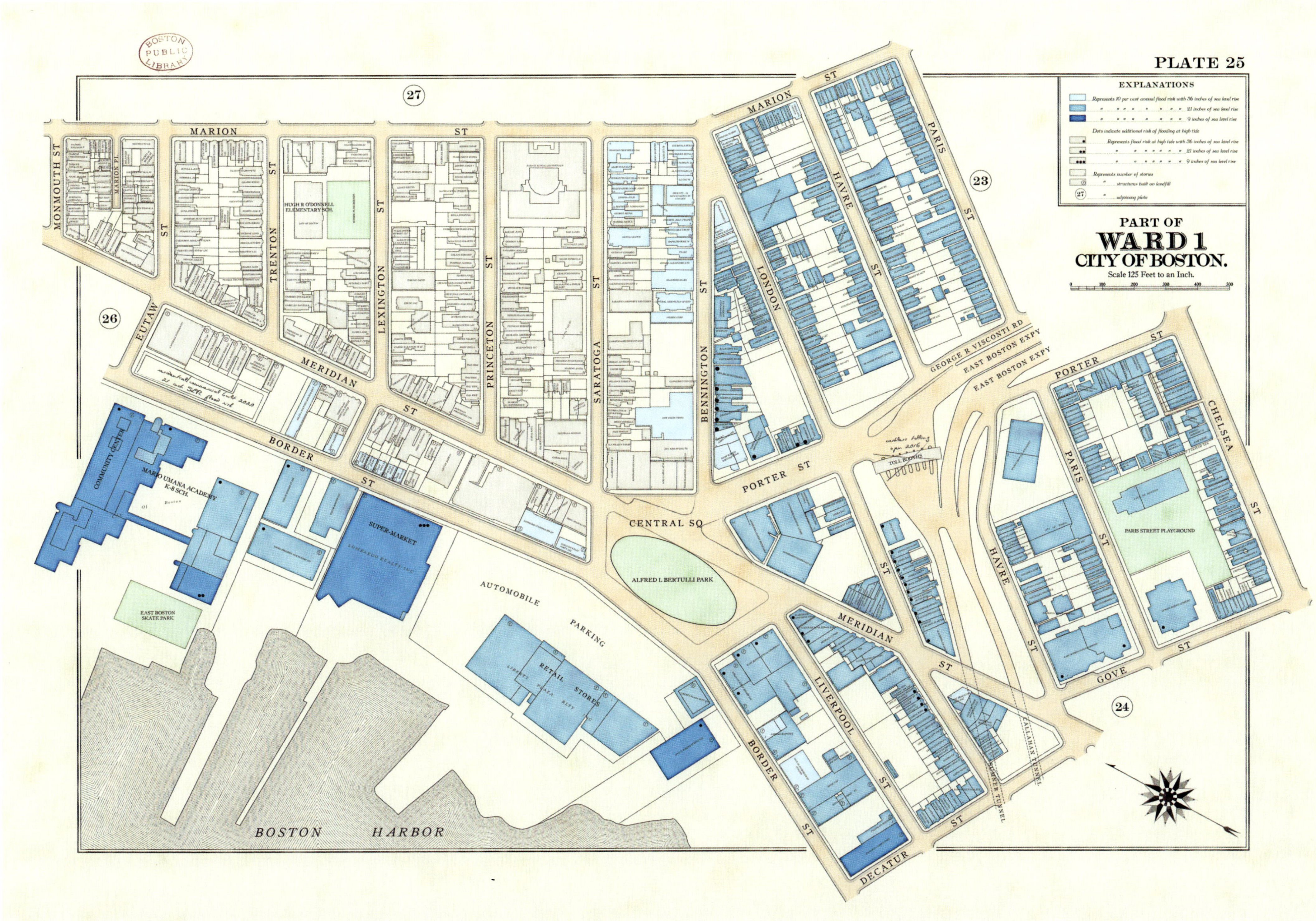
PLATE 25
EXPLANATIONS
Represents 10 per cent annual flood risk with 36 inches of sea level rise
" " " " " " " " 21 inches of sea level rise
" " " " " " " " 9 inches of sea level rise
Dots indicate additional risk of flooding at high tide
Represents flood risk at high tide with 36 inches of sea level rise
" " " " " " " 21 inches of sea level rise
" " " " " " " 9 inches of sea level rise
Represents number of stories
" structures built on landfill
27 " adjoining plate
PART OF
WARD 1
CITY OF BOSTON.
Scale 125 Feet to an Inch.
0 100 200 300 400 500
BOSTON PUBLIC LIBRARY
27
26
23
24
MARION ST
MONMOUTH ST
MARION PL
EUTAW ST
TRENTON ST
LEXINGTON ST
PRINCETON ST
SARATOGA ST
BENNINGTON ST
LONDON ST
HAVRE ST
PARIS ST
MERIDIAN ST
BORDER ST
PORTER ST
CENTRAL SQ
CHELSEA ST
GOVE ST
LIVERPOOL ST
DECATUR ST
GEORGE R VISCONTI RD
EAST BOSTON EXPY
CALLAHAN TUNNEL
SUMNER TUNNEL
TOLL BOOTHS
HUGH R O'DONNELL ELEMENTARY SCH.
COMMUNITY CENTER
MARIO UMANA ACADEMY K-8 SCH.
EAST BOSTON SKATE PARK
SUPER-MARKET
LOMBARDO REALTY, INC
AUTOMOBILE PARKING
RETAIL STORES
LIBERTY PLAZA REALTY INC
ALFRED L BERTULLI PARK
PARIS STREET PLAYGROUND
BOSTON HARBOR

CANADA
BRITISH COLUMBIA
ALBERTA
SASKATCHEWAN
MANITOBA
ONTARIO
QUEBEC
NEW BRUNSWICK
WASHINGTON
OREGON
CALIFORNIA
NEVADA
IDAHO
MONTANA
WYOMING
UTAH
COLORADO
ARIZONA
NEW MEXICO
NORTH DAKOTA
SOUTH DAKOTA
NEBRASKA
KANSAS
OKLAHOMA
TEXAS
MINNESOTA
IOWA
MISSOURI
ARKANSAS
LOUISIANA
WISCONSIN
ILLINOIS
MICHIGAN
INDIANA
OHIO
KENTUCKY
TENNESSEE
MISSISSIPPI
ALABAMA
GEORGIA
FLORIDA
SOUTH CAROLINA
NORTH CAROLINA
VIRGINIA
WEST VIRGINIA
PENNSYLVANIA
NEW YORK
NEW JERSEY
MAINE
PACIFIC OCEAN
ATLANTIC OCEAN
Gulf of Mexico
BAHAMAS
CUBA
LA HABANA
MÉXICO
SONORA
CHIHUAHUA
COAHUILA
TAMAULIPAS
Cd Victoria
RUSSIA
ALASKA
HAWAI'I
SEATTLE
SAN FRANCISCO
SAN JOSE
LOS ANGELES
SAN DIEGO
TIJUANA
LAS VEGAS
PHOENIX
DENVER
DALLAS
FORT WORTH
HOUSTON
SAN ANTONIO
CHICAGO
NEW YORK
PHILADELPHIA
LA GEOGRAFÍA ESENCIAL DE LOS ESTADOS UNIDOS DE AMÉRICA · LES ÉLÉMENTS FONDAMENTAUX DE LA GÉOGRAPHIE DES ÉTATS-UNIS D'AMÉRIQUE
THE ESSENTIAL GEOGRAPHY
of the United States of America · Second Edition
The United States of America and Outlying Areas

《 The Essential Geography of the United States of America, Second Edition, Version 4.4

Dave Imus, Imus Geographics

Geography is a primary means to appreciate our planet and its people, yet it remains on the margins of human collective consciousness.

This work offers a possible solution to this problem by introducing an art-centered form of cartography that disambiguates basic geography. It does this by making the patterns of land, water, and people, including nomenclature, immediately distinguishable and therefore simultaneously observable. This allows the viewer to see geographic relationships hidden by the visual discord of traditional wall maps.

Three years in development, a new terrain image reflects early-evening light, emphasizes form, and completes *The Essential Geography of the USA*, first published in 2010 as Edition 1.

The new terrain image draws inspiration from Canadian wildlife painter Robert Bateman, whose work directs the eye to form and space, in part by prioritizing broad strokes over small details. The resulting absence of small details in the terrain quiets visual chatter, making the overprinting lines and type far more accessible to the eye.

While scientifically faithful, this work arises from the art of illustration, rather than from the science of mapmaking. As such, it orchestrates graphic elements by employing a broad range of graphic principles, including minimalism, contrast, balance, visual layers, range of light, color theory, spacing, grouping, movement, continuity, unity, transparency, form, aerial perspective, realism, and reflectivity.

Critical to syntax, each line is rendered, and each type label is placed, entirely by eye.

Rigorous design and production in this illustration make geographic patterns and relationships observable, and basic geography intelligible.

Looking ahead, if cartography like this were incorporated into broader mapping systems, geography would move closer to the center of human collective consciousness, and public appreciation of our planet and its people would grow.

WASHINGTON
OREGON
IDAHO
COLUMBIA
OCEAN
Cascadia Basin
Gorda Basin
Gorda Ridges
SEATTLE
Space Needle
Univ of Washington
Boeing Aircraft Company
Pike Place Market
Ballard Locks
Redmond
Microsoft
PORTLAND
Forest Park
Washington Park
Powell's City of Books
Beaverton
Nike
Victoria
Saanich
Bellingham
Mt Baker 10,785
Anacortes
Burlington
Mt Vernon
Port Angeles
Port Townsend
Olympic NP
Olympic Mtns
7969
Olympic Peninsula
Shoreline
Everett
Marysville
Arlington
Snohomish
Edmonds
Kirkland
Redmond
Bellevue
Bremerton
Federal Way
Renton
Kent
Tacoma
Auburn
Puyallup
Olympia
Lacey
Hoquiam
Aberdeen
Ocean Shores
Grays Harbor
Willapa Bay
Long Beach
Cape Disappointment
Centralia
Chehalis
Raymond
South Bend
Mt Rainier 14,411
Mt Rainier NP
Wenatchee
Leavenworth
Ellensburg
Cle Elum
Yakima
Selah
Toppenish
Sunnyside
Grandview
Prosser
Richland
Kennewick
Pasco
Walla Walla
College Place
Moses Lake
Othello
Ephrata
Quincy
Ritzville
Spokane
Spokane Valley
Cheney
Coeur d'Alene
Hayden
Post Falls
Pullman
Moscow
Clarkston
Lewiston
Glacier Pk 10,541
Stevens Pass
Mt St Helens 8363
Mt Adams 12,276
Longview
Kelso
Vancouver
Camas
Columbia River Gorge NSA
Columbia
Snake
Columbia Basin
Hanford Reach NM
Hanford Nuclear Site
Okanogan
Omak
Oroville
Osoyoos
Colville
Kettle Falls
Sandpoint
Priest L
Pend Oreille
Astoria
Seaside
Tillamook
Lincoln City
Newport
Hillsboro
Beaverton
PORTLAND
Gresham
Lake Oswego
Oregon City
McMinnville
Salem
Keizer
Albany
Corvallis
Lebanon
Eugene
Springfield
Mt Hood 11,239
Hood River
The Dalles
Mt Jefferson 10,497
Bend
Redmond
Prineville
Madras
South Sister 10,358
Three Sisters Wldns
Newberry NVM
Paulina Pk 7984
Florence
Oregon Dunes NRA
Coos Bay
North Bend
Roseburg
Grants Pass
Medford
Ashland
Mt McLoughlin 9495
Crater Lake NP
Crater L
Mt Thielsen 9182
Klamath Falls
Upper Klamath L
Gearhart Mtn 8370
Lakeview
Brookings
Crescent City
Redwood NP
Yreka
Mt Shasta 14,162
Mt Eddy 9025
Lava Beds NM
Tule Lake Internment Camp
Goose L
Pendleton
Hermiston
Umatilla
La Grande
Baker City
Wallowa Mtns
Elkhorn Rng
Blue Mtns
John Day
John Day Fossil Beds NM
Painted Hills Unit
Ochoco Mtns
Strawberry Mtns 9038
Burns
Hines
Malheur L
Malheur NWR
Steens Mtn
Alvord Desert
Hart Mtn NAR
Pueblo Mtns 8725
Wagontire Mtn 6513
Fort Rock
Ontario
Nyssa
Payette
Fruitland
Weiser
Boise
Meridian
Nampa
Caldwell
Kuna
Eagle
Emmett
Mountain Home
Twin Falls
Jerome
Hells Canyon NRA
He Devil 9393
Seven Devils
Grangeville
Salmon
Salmon River Mtns
Frank Church-River of No Return Wldns
Sawtooth Rng
Sawtooth NRA
Borah Pk 12,662
Hyndman Pk 12,009
Ketchum
Sun Valley
Hailey
Craters of the Moon NM & P
Snake River Birds of Prey NCA
Hagerman Fossil Beds NM
Minidoka Internment NM
Sheldon NWR
Bitterroot Rng
Selway-Bitterroot Wldns
Clearwater Mtns
Dworshak Res
Cabinet Mtns

ALBERTA
SASKATCHEWAN
MONTANA
NORTH DAKOTA
SOUTH DAKOTA
WYOMING
Lethbridge
3048
Medicine Hat
2198
Redcliff
Fort Macleod
Taber
Bow Island
Raymond
Magrath
Cardston
Claresholm
Vauxhall
Picture Butte
Milk River
Cypress Hills
4567
4810
Fox Valley
Swift Current
2680
Gull Lake
Maple Creek
Shaunavon
Val Marie
GRASSLANDS NP
Wood Mtn
3350
Gravelbourg
Assiniboia
Willow Bunch
Moose Jaw
Lumsden
RCMP Academy
Regina
1893
Indian Head
Fort Qu'Appelle
Qu'Appelle
Grenfell
Old Wives L
L Diefenbaker
Strasbourg
Melville
Yorkton
Langenburg
Esterhazy
Whitewood
Moosomin
Weyburn
Radville
Stoughton
Carlyle
Estevan
Oxbow
Carnduff
Roblin
Dauphin
2729
968
Grandview
RIDING MTN NP
Ste Rose du Lac
Lake Manitoba
Wasagaming
2480
Minnedosa
Neepawa
Portage la
Rivers
Carberry
Virden
Brandon
1343
Souris
Assiniboine
Melita
Boissevain
Deloraine
2516
Killarney
Turtle Mtn Int'l Peace Garden
Hollar
Cleveland
9466
BLACKFEET RES
Triple Divide Pk
8020
Browning
4377
Sunburst
Mt Royal
6914
Cut Bank
Shelby
Chester
Valier
Conrad
Two Medicine
Badger
Marias
Milk
ROCKY BOYS RES
Big Sandy
Havre
2491
Chinook
Harlem
6900
Bear Paw Battlefield–Nez Perce NHP
FORT BELKNAP RES
Malta
Saco
Teton
Choteau
Dutton
Fairfield
Vaughn
Fort Benton
Geraldine
Great Falls
3312
Belt
Sun
Cascade
Upper Missouri River Breaks NM
Charles M Russell NWR
Glasgow
2095
Nashua
Wolf Point
Poplar
FORT PECK RES
2685
Scobey
Plentywood
Culbertson
Fort Peck
Fort Peck Dam
Missouri
Fort Peck Res
2250
Jordan
Circle
Richey
Sidney
Fairview
Stanford
Judith Basin
Lewistown
3963
Little Belt Mtns
9175
Winnett
Big Snowy Mtns
8730
Judith Gap
Harlowton
Ryegate
Roundup
Melstone
Musselshell
White Sulphur Sprs
Townsend
Helena
4058
Gates of the Mtns
Big Belt Mtns
Canyon Ferry L
9505
Boulder
Butte
Berkeley Pit
5538
Whitehall
Headwaters of Missouri River
Three Forks
Belgrade
Bozeman
4795
Livingston
Bridger Rng
9573
Crazy Mtns
11,230
Big Timber
Columbus
Billings
3117
Laurel
Bull Mtns
4701
Pompeys Pillar NM
Hysham
Forsyth
Miles City
2371
Yellowstone
Terry
Glendive
2069
Wibaux
Big Sheep Mtn
3386
Theodore Roosevelt NP Elkhorn Ranch Site
Theodore Roosevelt NP South Unit
Theodore Roosevelt NP North Unit
Baker
2936
Ekalaka
Colstrip
Tongue
Hardin
Crow Agency
3038
CROW RES
Little Bighorn Battlefield NM
NORTHERN CHEYENNE RES
Lodge Grass
Broadus
Powder
4206
Absaroka-Beartooth Wldns
Granite Pk
12,799
Red Lodge
Bridger
Bighorn Cyn NRA
Ennis
Sheridan
Virginia City
Dillon
Ruby Rng
Madison Rng
Gallatin
Big Sky
11,316
West Yellowstone
Monida Pass
6870
Mammoth
YELLOWSTONE NAT'L PARK
Old Faithful
Yellowstone L
7729
Absaroka Range
Beartooth Pass
10,947
Cowley
Byron
Lovell
Powell
Medicine Wheel
Dayton
Ranchester
Sheridan
3745
Trout Pk
12,244
Cody
5016
Heart Mtn Internment Camp
Greybull
Basin
Bighorn Basin
Meeteetse
13,167
Cloud Pk
Bighorn Mtns
Story
Buffalo
Worland
4061
Francs Pk
13,153
Togwotee Pass
Thermopolis
Boysen Res
4800
Shoshoni
WIND RIVER RES
Gannett Pk
13,804
Dubois
Jackson
6208
Jackson L
6772
GRAND TETON NP
Teton Rng
13,770
Rexburg
Driggs
Idaho Falls
4710
Shelley
Blackfoot
Ammon
Rigby
St Anthony
Ashton
FORT HALL RES
Alpine
Freedom
Clearmont
Gillette
4544
Kaycee
Hole-in-the-Wall
Midwest
Teapot Dome
Wright
Powder River Basin
Moorcroft
Upton
Devils Tower NM (Mato Tipila)
Hulett
Sundance
Belle Fourche
Spearfish
Lead
Deadwood
Sturgis
3422
Black Hills Motor Classic
Newcastle
Osage
Black Elk Pk
7242
Black Hills
Crazy Horse Mem
Mt Rushmore Nat'l Mem
Custer
Custer SP
Jewel Cave NM
Wind Cave NP
Rapid City
3247
Box Elder
Wall Drug
Wall
Minuteman Missile NHS
BADLANDS NP
Kadoka
Ghost Dance
Philip
Cheyenne
Crosby
1986
Portal
Northgate
Bowbells
Kenmare
Mohall
Tioga
Ray
Stanley
Williston
1882
Lake Sakakawea
1850
New Town
Berthold
Minot
1621
Souris
Turtle Mtns
Bottineau
Dunseith
Rolla
TURTLE MTN RES
Rolette
Rugby
Cando
Langdon
Devils Lak
1475
Velva
Fort Buford
Watford City
FORT BERTHOLD RES
Parshall
Garrison
Harvey
SPIRIT LK RES
New Rockford
Killdeer Mtns
3314
Killdeer
Garrison Dam
Hazen
Underwood
Washburn
Beulah
Carrington
Hawks Nest
2121
Wilton
Medora
Belfield
Dickinson
2417
Richardton
Badlands
New Salem
Glen Ullin
Mandan
Bismarck
1700
Int'l Powwow
Lincoln
Steele
Jamestown
1419
Fort Abraham Lincoln SP
New England
Cannonball
Elgin
White Butte
3506
Mott
Bowman
Pretty Rock Butte
2799
Hettinger
Napoleon
Linton
Wishek
Edgeley
1875
Ashley
Ellendale
Fort Yates
Lemmon
2569
McIntosh
STANDING ROCK RES
McLaughlin
Eureka
Grand
Buffalo
Slim Buttes
3671
Bison
Mobridge
Selby
Ipswich
Bowdle
Aberdeen
1300
L Oahe
1608
CHEYENNE RIVER RES
Eagle Butte
2418
Faith
Newell
Gettysburg
Faulkton
Redfield
Onida
Pierre
1484
Oahe Dam
Highmore
Miller
Fort Pierre
LOWER BRULE RES
CROW CREEK RES
1996
Fort Thompson
Murdo
White River
Chamberlain
Kimball
White
1368
Little Missouri
Yellowstone
Madison
Jefferson
Bighorn
Wind
Snake
Belle Fourche
Cheyenne

Logan Glacier
Walsh Glacie
Logan Glacier
Baldwin Glacier
Ogilvie Gl
CANADA
UNITED STATES
Queen Pk
17651
Prospector Peak
18517
Mount Logan
19551
Catenary Peak
13442
McArthur Peak
14252
King Peak
16972
Teddy Pk
12979
acier
Quintino Sella Glacier
Mount Logan has the
circumference of any
mountain on Earth. Th
plateau is 10 miles lon
cap nearly 1,000 feet t
5945
6299
Seward Glac
ley
5900
Columbus Glacier
6693
Mount Saint Elias
(Was'eitushaa)
18008
Mount Newton
13045
Mount Jeannette
11601
Mount Bering
11496
Mount Malaspina
12490
Mount Augusta
14070
Mount Eaton
10587
Mt Huxley
12067
The Hump
12388
Haydon Peak
11945
Corwin Cliffs
4649
Newton Glacier
Crater Basin
Cascade Glacier
Augusta Gl
Seward Glacier
Mount Owen
7874
Yahtse Glacier
Tyndall Glacier
Libbey Glacier
Agassiz Glacier
4787
Samovar Hills
Pinnacle Pass
Guyot Hills
4905
Karr Hills
3205
2300
Hitchcock Hills
1957 glacier extent
1972
Taan Fiord
Moore Nunatak
2064
3773
Chaix Hills
Tsaa Fiord
1957
1950
Kageet Point
Kichyatt Point
At 40 miles wide and up to 2,000 feet thick, Malaspina Glacier is a textbook example of a piedmont glacier. Rubble, meltwater, and vegetation cover its outermost fringes. The glacier is named after Alessandro Malaspina, an Italian explorer sailing for Spain in 1791.
1650
Icy Bay
Gull I
Malaspina Glac
(Sít' Tlein)
1904
Tsimpshian Point
Riou Bay
The Saint Elias Mountains are the highest coastal range on Earth. Only 11 miles separate uppermost Taan Fiord from the 18,008-foot summit of Mount Saint Elias. Further inland, 19,551-foot Mount Logan has the distinction of being the highest peak in Canada and second highest peak in North America.
Sitkagi Bluffs
Cape Sitkagi

The Icefield Ranges are nunataks, a Greenlandic term for "island" mountains surrounded by glaciers.
ld Ranges
Kaskawulsh Glacier
Kaskawulsh Mtn
9747
Mount Leacock
10148
Stairway Glacier
South Arm
7664
Mount Queen Mary
12887
6286
Mount King George
12274
Pinnacle Peak
12070
Disappointment Gl
Hubbard Glacier
Hubbard Glacier
Mount Alverstone
14500
Mount Hubbard
14951
Mount Kennedy
13944
Dusty
Mount Vancouver
15787
4593
5906
YUKON
ALASKA
Alverstone Glacier
The Weisshorn
11624
ount Irving
4200
Ulu Mountain
10102
Cathedral Glacier
Mount Foresta
11001
Icefield
Mount Cook
13766
4987
Mount Seattle
10350
Hubbard Glacier
3000
Valerie Glacier
Mount Jetté
8425
Haenke Glacier
Turner Glacier
Hubbard Glacier
BRITISH COLUMBIA
ALASKA
Blossom Gl
Hayden Glacier
Lucia Glacier
Variegated Glacier
Mount Stamy
4557
Orange Glacier
Gilbert Point
3804
Atrevida Glacier
Galiano Gl
Bancas Point
Disenchantment Bay
Haenke Island
Butler Glacier
Floral Hills
Mt Alexander
3805
Russell
Mt McCarty
3519
Calahonda Cr
Point Latouche
Osier Bluffs
Cape Enchantment
Nunatak Fiord
Mount Draper
5643
Grand Wash River
Mt Hendrickson
4590
Blizhni Point
Fiord
Malaspina Lake
Surging Hubbard Glacier occasionally chokes off the entrance to Russell Fiord. The increasingly fresh water imperils marine mammals trapped in the fiord. If the water were to rise too high, it could overflow the southern bank and drain through the Situk River.
Seal Bay
Mount Ruhamah
5520
Mount Tebenkof
4219
Mount Pinta
5530
Hidden Glac
Mt Mallott
3460
Knight Island
Tia-Xagh Island
Yakutat Bay
Krutoi Island
Lake Redfield
Situk Lake
Mountain Lake
Northeast Point
Kriwoi Island

Malaspina Glacier and the Saint Elias Mountains »

Tom Patterson, US National Park Service (retired)

High, wild, remote, and icy—these are some of the words that describe the Saint Elias Mountains on the border of Alaska and Yukon. They are the highest coastal mountains on Earth, including the second- and fourth-highest peaks in North America (Logan and St. Elias, respectively). Massive tidewater glaciers spill down their slopes. And sprawling Malaspina Glacier, some 40 miles wide and 2,000 feet thick, is a textbook example of a piedmont glacier.

However, the Saint Elias Mountains are little known because of sparse population, limited road access, and an international border that divides them. The available maps reflect this dichotomy. For example, the official map of Wrangell–St. Elias National Park focuses on the US side of the border, as do Canadian maps for their territory. Another problem: the Wrangell–St. Elias National Park map emphasizes administrative boundaries more than mountains and glaciers.

As a remedy, I rendered the Malaspina Glacier panorama to showcase a wild landscape in its entirety, where human development is minimal. Malaspina Glacier, with its concentric rings of ice, rubble, and meltwater, is front and center. The glistening Saint Elias Mountains serve as a backdrop that contrasts with the cold waters of the Pacific Ocean in the foreground. Given the emphasis on natural features, the international border appropriately appears inconspicuously on the landscape.

My panorama initially featured a curved horizon and a sky decorated with wispy clouds that mimicked the classic panorama style popularized by Heinrich Berann in the mid-twentieth century. However, I was concerned that the sky and unessential background land occupied too much space, detracting from the more important foreground. Cropping out the sky and filling the background with a dark gradient resulted in a better overall composition and more efficient use of space; further, the darkened background added ideal contrast for placing the title.

I started the project in 2017 and then put it aside for four years. However, accelerating climate change brought newfound urgency to my mapping. For example, nearby Icy Bay has experienced 31 miles of glacier retreat since 1904. I wanted to map beautiful Malaspina Glacier while it still existed.

Malaspina Glacier and the Saint Elias Mountains

Alaska / Yukon / British Columbia

MAP LEGEND
Prehistoric Site
Ancient Findings
Venetian Castle
Venetian Tower
Museum
Imp/ant Building
Industry Building
Winery
Windmill
Carved Dwellings
Pumice Heart
Pumice Quarry
Active Volcano
Bicycle Tours
Open Air Cinema
Art Gallery
Bookstore
Jewellery
Market/Shop
Trailhead
Gas Station
Bus Station
Taxi Station
Parking
Post Office
Cash Machine
Monastery
Church/Chapel
Port
Sailboat Port
Lighthouse
Sea Kayak
Swimming Place
Hot Waters
Sunset View
Tomato
Caper
Split Pea
Traditional Vine
Pharmacy
Vineyards
Settlement
Hiking Route
Main Asphalt Road
Asphalt Road
Dirt Road
Stairsteps
Footway
VINEYARDS
Santorini's fabulous Wines are produced with a peculiar way.
Due to the strong Aegean winds, the vineyards are grown in a circular shape to be protected from the winds and the drought.
LOCAL DELICACIES
TOMATOES
Tiny cute and extremely delicious tomatoes are some of Santorini's Local Products.
CAPER
Caper is growing and cultivated everywhere on the island.
SPLIT PEAS
Split Peas or Fava, as they are called in Greece, is a very common local product of Santorini.
HIKING ROUTES
1. Pyrgos - Profitis Ilias Mon. - Ancient Thera
2. Pyrgos - Emboreio
3. Kamari - Ancient Thera - Perissa
4. Pyrgos - Agios Georgios - Panagia Episkopis
5. Pyrgos - Exo Gonia - Mesa Gonia - Pyrgos
6. Pyrgos - Megalochori
7. Megalochori - Stavros - Plaka
8. Pyrgos - Vothonas - Panagia Sergena
9. Fira - Imerovigli - Oia
10. Foinikia - Panagia - Koloumbos
11. Apothikes - Balos - Akrotiri
12. Akrotiri - Kokkini Paralia - Taxiarchis - Faros
13. Akrotiri - Agia Anna - Theros - Vlychada
14. Oia - Ammoudi
15. Oia - Armeni
16. Imerovigli - Skaros - Panagia Theoskepasti
17. Epano Fira - Kato Fira
18. Megalochori - Stavros - Thermia
19. Diapla
20. Ancient Thera
21. Manolas - Potamos - Agrilia - Manolas
22. Manolas - Profitis Ilias - Kera - Panagia Mon.
23. Agrilia - Christos - Pumice Quarry
24. Korfos - Manolas
25. Volcano (Tholos Nautilos)
SANTORINI
Santorini is a complex of five separate islands, namely Thera, Therasia, Aspronisi, Palaia Kameni and Nea Kameni.
All these five islands are volcanic. They were built piece by piece from numerous volcanic eruptions that occurred in the past.
CALDERA
The great circular gulf in the middle of Santorini is a Caldera, a collapsed enormous former volcanic crater.
PREHISTORIC AKROTIRI
The oldest civilization on the island, it was destroyed from the notorious Minoan volcanic eruption in 1613 B.D. A must see archaeological site.
ANCIENT THERA
After the Minoan catastrophe the island remained uninhabited for many hundreds of years until approximately 1000 BD, when ancient Greeks occupied it.
THERA isl.
THERASSIA isl.
NEA KAMENI isl.
PALAIA KAMENI isl.
ASPRONISI isl.
OIA
FIRA
1:40,000
1:7,500
Map Edition: 2024
Map Design: Staridas Geography
Copyright 2024 Staridas Geography
All Rights Reserved / CC BY-NC-ND 4.0

« Santorini Islands

Spyridon (Spiros) Staridas

This is a map of the Santorini insular complex, located at the south of the Cyclades Archipelago, in the middle of the Aegean Sea, Greece. White chapels with blue rooftops, prehistoric archaeological sites, medieval castles and monasteries, and a breathtaking view of an active volcano are just some of the elements of Santorini's rich heritage. Santorini also offers high-quality local produce and some of the most exquisite wines in Europe, which are registered with a protected designation of origin (PDO).

The island's beauty is rich yet simple, and I wanted to depict that essence on this map. I used an uncluttered background, smooth pastel colors, and a pictorial (but not childish) touch on the symbology to attract the reader's eye long enough for them to read the map and learn from it.

A limited color palette was chosen, with light brown for the land and light green for vineyards, while man-made areas, like settlements, are shown in darker brown. A light cyan vignette was added along the coastline to enhance the feeling of being on an island.

The point symbols are hand-drawn, and I did a meticulous study of the island's points of interest to collect their locations, names, and characteristics so I could decide on intuitive icons for the reader. These fall into two major categories, one including simpler icons with a solid color and one with slightly more complex multicolored icons, and were created with the help of Margarita Kondylaki, a talented graphic designer. All of them have a more or less organic shape to contribute to the impression of a hand-drawn map, which is further enhanced by a subtle watercolor paper texture.

At the bottom left, there is a large-scale map of the settlement of Oia, definitely a must-visit for those who come to the island, with one of the most famous and scenic sunsets in the Mediterranean Sea.

Finally, there are suggested hiking trails, guiding visitors to some of the unique and unexpected sights they might otherwise miss, along with short explanatory text to further inform readers of the island's sights and experiences.

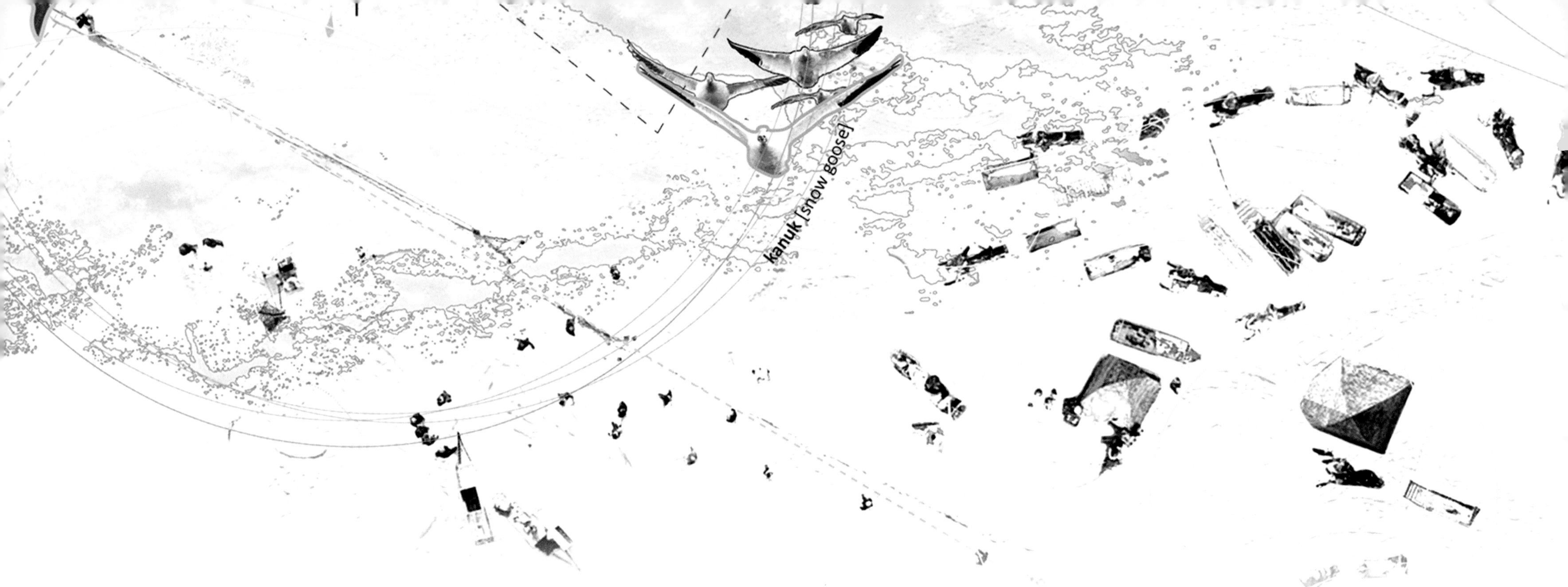

The Bowhead Hunt as an Inupiaq Calendar System »

Amanda Aman

In Inupiaq culture, the bowhead whale is a keystone for physical and spiritual sustenance. The hunt is a system of ephemeral infrastructure that essentially defines the Inupiaq calendar and involves the entire community throughout the course of the year. The carving of shore fast sea ice trails in the fall initiates the process, as community members use snow machines (that is, snowmobiles) and ice picks to carve safe trails along frozen coastlines that essentially extend land into the ocean. Exact trail locations vary each year, and the trails are constructed with an intimate knowledge of a number of natural systems: bowhead migration routes, the effect of wind on snowdrifts, and weather patterns. There is also an elaborate vocabulary about the formation and type of ice, one that also describes its sentience and agency, so as to be completely in tune with both what is seen and what is hidden. Camps are established along the ice edge, and the hunt begins. These camp networks are then used in the spring to hunt migrating waterfowl, using the same intimate knowledge of place and its natural systems.

The map was constructed by collapsing time and space together, creating a palimpsest of seasonal sequences and spatial operations. The peripheral loop details the sequence of the annual hunt while spatially correlating to the imaginative projection of the event in the center. Waterfowl migrations are woven into this projection as well, aligning with predominant wind patterns that inform their aerial routes in the spring after having formed terrestrial routes for hunters in the fall. The reader of the map can begin to imagine highways of whales and waterfowl, looped over and through changing trails of human mobility, to remake geography and reality in ways that speak to the Inupiaq way of life. In contrast to the traditional mapping of physical geography, these realities are never remade the same, are always changing, and are always in dialogue with one another. Realizing them in this way lays the groundwork for advocacy by communicating the importance of ephemeral systems as it relates to Inupiaq livelihood.

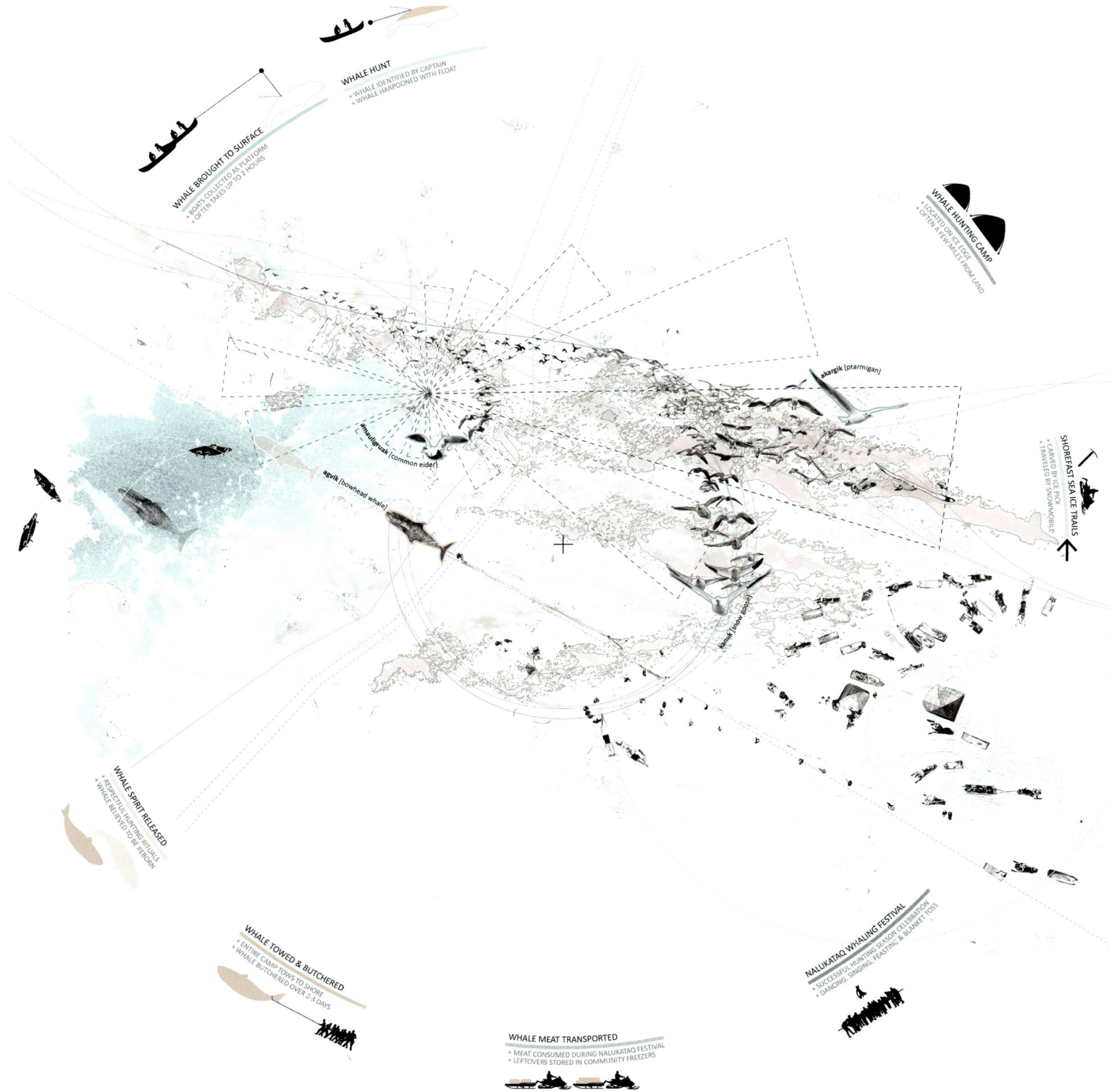

WHALE HUNT
+ WHALE IDENTIFIED BY CAPTAIN
+ WHALE HARPOONED WITH FLOAT
WHALE BROUGHT TO SURFACE
+ BOATS COLLECTED AS PLATFORM
+ OFTEN TAKES UP TO 2 HOURS
WHALE HUNTING CAMP
+ LOCATED ON ICE EDGE
+ OFTEN A FEW MILES FROM LAND
akargik [ptarmigan]
amauligruak [common eider]
agvik [bowhead whale]
kanuk [snow goose]
SHOREFAST SEA ICE TRAILS
+ CARVED BY ICE PICK
+ TRAVELED BY SNOWMOBILE
WHALE SPIRIT RELEASED
+ RESPECTFUL HUNTING RITUALS
+ WHALE BELIEVED TO BE REBORN
WHALE TOWED & BUTCHERED
+ ENTIRE CAMP TOWS TO SHORE
+ WHALE BUTCHERED OVER 2-3 DAYS
WHALE MEAT TRANSPORTED
+ MEAT CONSUMED DURING NALUKATAQ FESTIVAL
+ LEFTOVERS STORED IN COMMUNITY FREEZERS
NALUKATAQ WHALING FESTIVAL
+ SUCCESSFUL HUNTING SEASON CELEBRATION
+ DANCING, SINGING, FEASTING & BLANKET TOSS

Wien Secession, an Artsy Walk through Vienna »

Camila Narbaitz Sarsur

Paper maps never lose their charm, and walking through a city with a paper map can be like reading a paperback novel. However, there's always the opportunity to push their utility a bit further. This guide-map explores how paper maps can show more than just the "must see" landmarks of a city. It does so by including a thematic narrative—unraveling the cultural, architectural, and social evolution of Vienna through its Secessionist movement and its lasting influence on the city's modern identity. To achieve this, three main approaches were taken. First, the simple yet ingenious mechanic of folding allows the story behind the map to be revealed like the narrative in a book. Each unfolding unveils a layer of the Secessionist movement's story. Second, it offers more than mere points on a map: a timeline and introduction present the movement and delve into historical facts. Additionally, a relational graph introduces the movement's inspiration and its members, explores their relationships, and shows their creations, which appear on the main map. Third and last, the simple map, with a single folding action, can be overlaid with a suggested schematic route located in the four corners of the guide. The decision not to pre-draw routes on the map permits the user to make their own way through the city.

The color palette, textures, fonts, and symbology of this map are inspired by the graphic design of the Secessionist movement. Its distinctive style has been emulated through the use of geometric forms balanced with organic elements, a color scheme of earth tones and pastels, and the incorporation of gold details for accent and elegance. The overall design strikes a balance between simplicity and ornamentation, reflecting the Secessionist emphasis on refined craftsmanship and attention to detail. All the data layers that composed the map were taken from Open Data Österreich. The main source for the historical information was the book *Fin-De-Siècle Vienna* by Carl E. Schorske (first published in 1979).

A FOLDABLE GUIDE MAP

to travel across space and time, with a spirit of art and adventure.

This is not a guide for art or city experts. This is a guide for those who want to understand Vienna by discovering how its artistic, social and cultural movements in 1900 shaped its urban fabric.

This guide contains: 1 map and 1 guide of facts, relationships and suggestions to delve into Vienna's Secession movement.

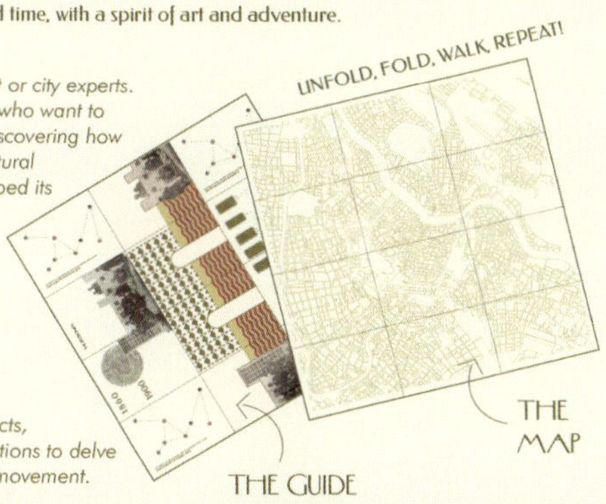

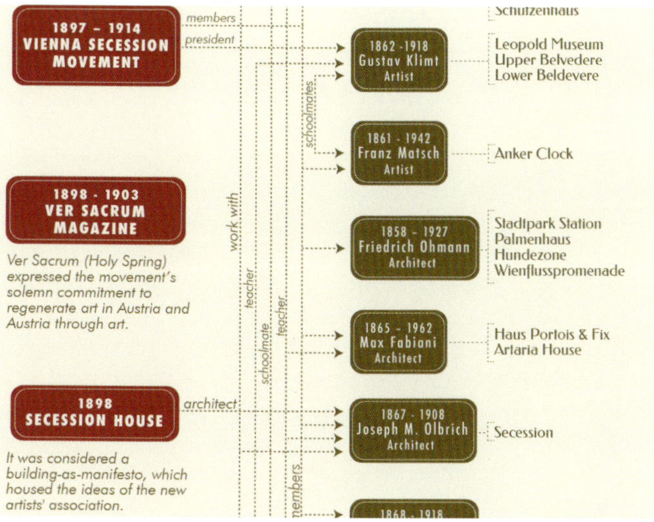

HOW DOES IT WORK?

After reading this introduction, keep on unfolding the map and discover history, main characters and inspirations behind the Vienna Secession movement.

On the corners of the guide-map, you will find 4 different suggested routes. With a single folding action, the chosen schematic route can be followed by using the underneath map.

This map was created, reading "Fin-De-Siecle Vienna" by Carl E. Schorske as the main source.

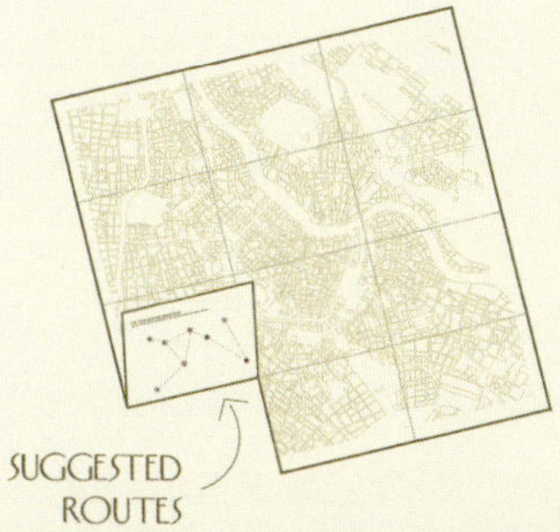

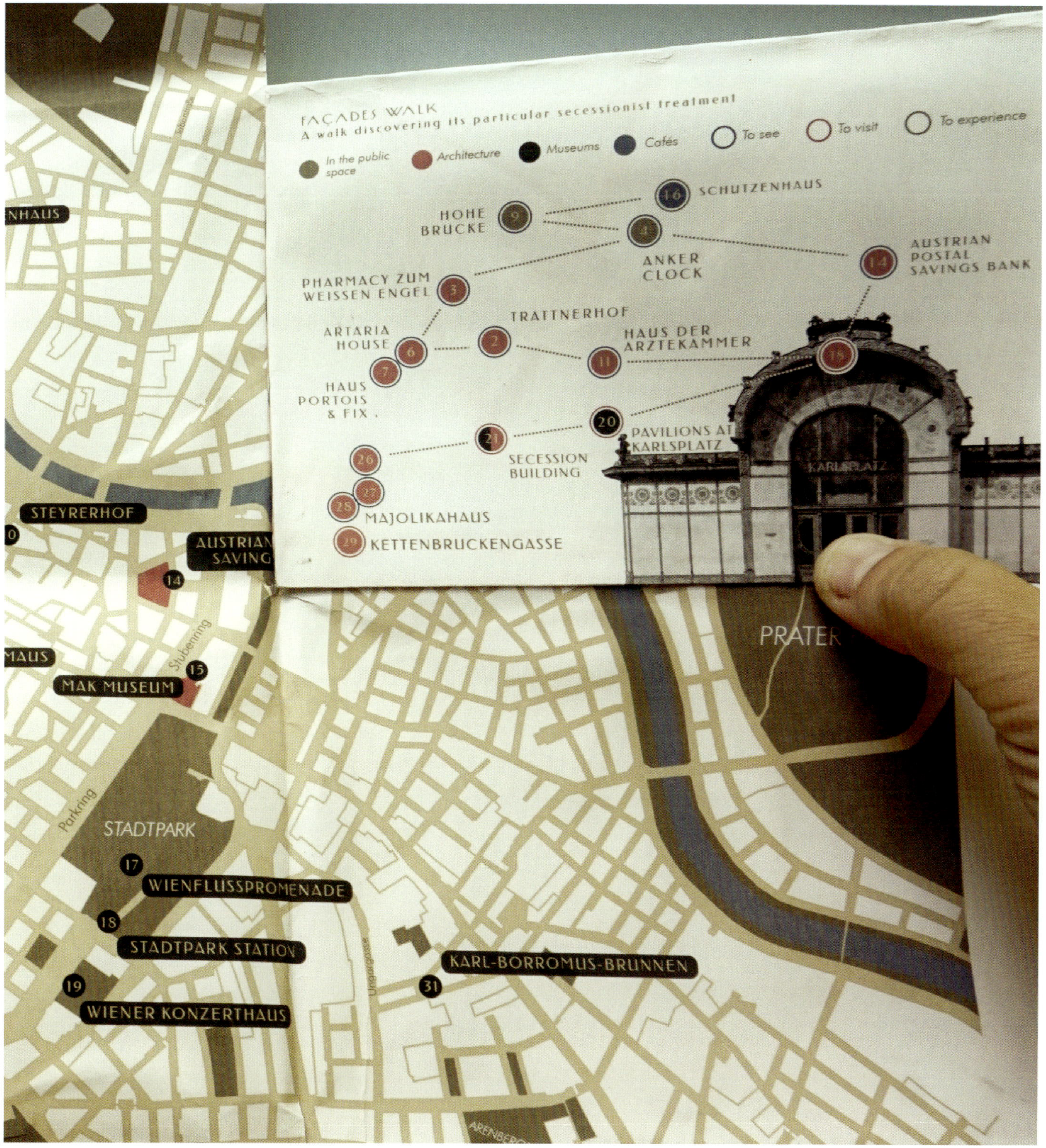

DANUBE
AUGARTEN
STRUDLHOF STAIRWAY
SIGMUND FREUD MUSEUM
SCHUTZENHAUS
HOHE BRUCKE
ANKER CLOCK
STEYRERHOF
AUSTRIAN POSTAL SAVINGS BANK
CAFE CENTRAL
CAFE KORB
CABARET FLEDERMAUS
MAK MUSEUM
PRATER PARK
VOLKSGARTEN
HELDENPLATZ
PALMENHAUS
BURGGARTEN
STADTPARK
WIENFLUSSPROMENADE
LEOPOLD MUSEUM
STADTPARK STATION
KARL-BORROMUS-BRUNNEN
SECESSION BUILDING
WIENER KONZERTHAUS
FILLGRADERSTIEGE
CAFE SPERL
PAVILIONS AT KARLSPLATZ
LINKE WIENZEILE
MAJOLIKAHAUS
LOWER BELDEVERE MUSEUM
KETTENBRUCKENGASSE
CAFÉ RUDIGERHOF
BELVEDERE GARTEN
UPPER BELVEDERE MUSEUM
REFERENCES
2. Trattnerhof
3. Pharmacy Zum Weissen Engel
6. Artaria House
7. Haus Portois & Fix
8. Wiener Werkstate
11. Haus der Arztekammer
26 & 27. Linke Wienzeile

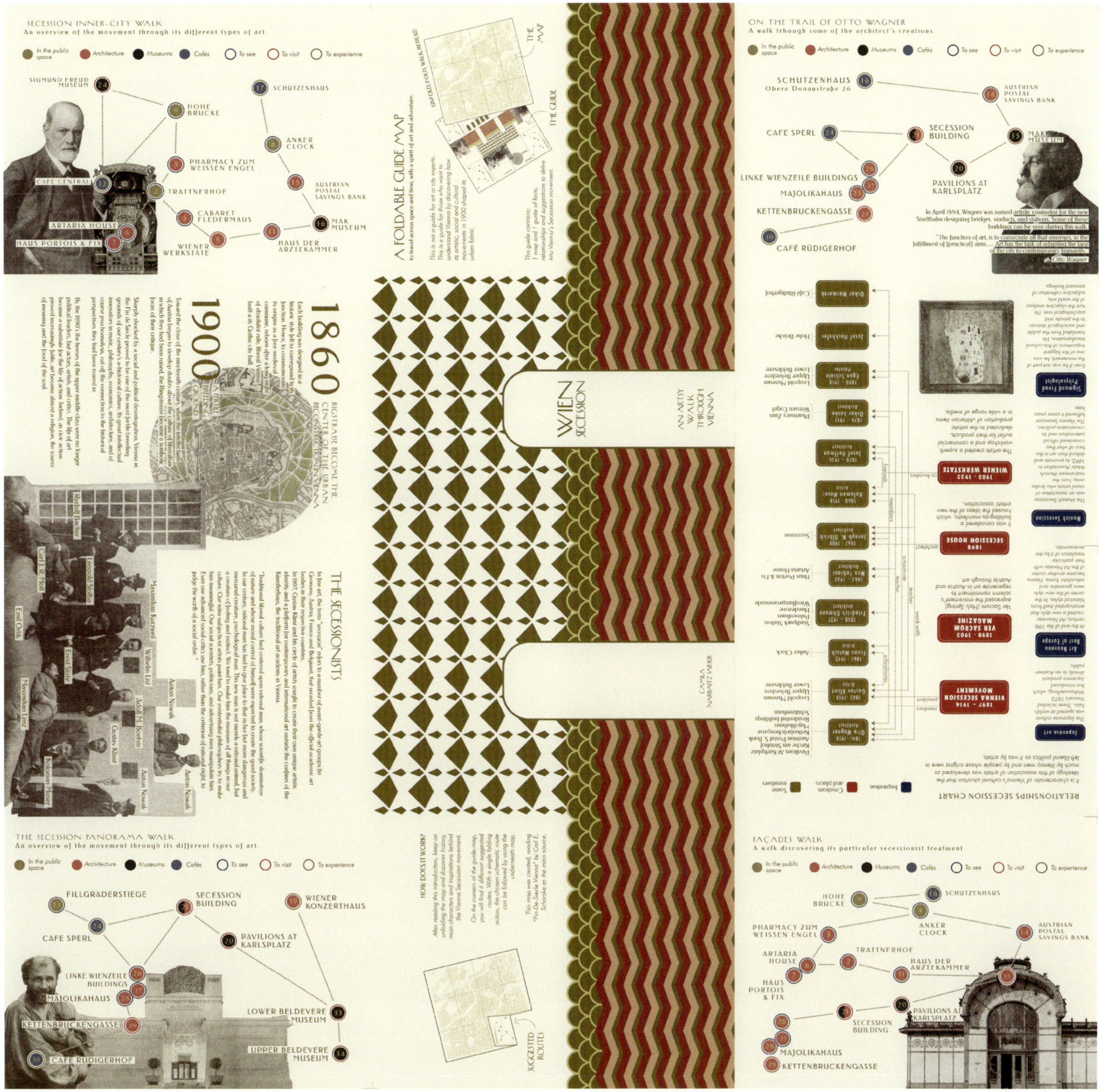
SECESSION INNER-CITY WALK
An overview of the movement through its different types of art.
In the public space
Architecture
Museums
Cafés
To see
To visit
To experience
SIGMUND FREUD MUSEUM
HOHE BRUCKE
SCHUTZENHAUS
ANKER CLOCK
CAFE CENTRAL
PHARMACY ZUM WEISSEN ENGEL
TRATTNERHOF
AUSTRIAN POSTAL SAVINGS BANK
CABARET FLEDERMAUS
ARTARIA HOUSE
MAK MUSEUM
HAUS PORTOIS & FIX
WIENER WERKSTATE
HAUS DER ARZTEKAMMER
A FOLDABLE GUIDE MAP
THE MAP
THE GUIDE
WIEN SECESSION
AN ARTSY WALK THROUGH VIENNA
1860
1900
THE SECESSIONISTS
ON THE TRAIL OF OTTO WAGNER
A walk trhough some of the architect's creations.
SCHUTZENHAUS
Obere Donaustraße 26
AUSTRIAN POSTAL SAVINGS BANK
CAFE SPERL
SECESSION BUILDING
MAK MUSEUM
LINKE WIENZEILE BUILDINGS
PAVILIONS AT KARLSPLATZ
MAJOLIKAHAUS
KETTENBRUCKENGASSE
CAFÉ RÜDIGERHOF
In April 1894, Wagner was named artistic counselor for the new Stadtbahn designing bridges, viaducts, and stations. Some of these buildings can be seen during this walk.
"The function of art, is to consecrate all that emerges, in the fulfillment of [practical] aims... Art has the task of adapting the face of the city to contemporary humanity."
Otto Wagner
RELATIONSHIPS SECESSION CHART
THE SECESSION PANORAMA WALK
An overview of the movement through its different types of art.
FILLGRADERSTIEGE
SECESSION BUILDING
WIENER KONZERTHAUS
CAFE SPERL
PAVILIONS AT KARLSPLATZ
LINKE WIENZEILE BUILDINGS
MAJOLIKAHAUS
KETTENBRUCKENGASSE
LOWER BELDEVERE MUSEUM
UPPER BELDEVERE MUSEUM
CAFE RUDIGERHOF
HOW DOES IT WORK?
SUGGESTED ROUTES
FAÇADES WALK
A walk discovering its particular secessionist treatment
HOHE BRUCKE
SCHUTZENHAUS
PHARMACY ZUM WEISSEN ENGEL
ANKER CLOCK
AUSTRIAN POSTAL SAVINGS BANK
ARTARIA HOUSE
TRATTNERHOF
HAUS DER ARZTEKAMMER
HAUS PORTOIS & FIX
SECESSION BUILDING
PAVILIONS AT KARLSPLATZ
MAJOLIKAHAUS
KETTENBRUCKENGASSE

红莲湖景区
885水世界
Linzi North Railway Station
临淄北站
Zibo North Railway Station
淄博北站
桐林遗址
路家山
齐故城排水道口
土山
凤凰山
晏婴墓
淄博高新北收费站
傅山自然地质博物馆
高新区（开发区）
临淄南金农民博物馆
巧媳妇酱文化产业园
稷下学宫遗址
齐国故城遗址博物馆
西天寺造像
张
店
区
临
淄
区
青银高速淄博服务区
淄博收费站
淄博高新区收费站
淄博烧烤海月龙宫体验地
海岱楼钟书阁
华光国瓷艺术中心
鲁泰快速路
烧烤专线 1,2,3 (高新区)
淄博西收费站
中国陶瓷琉璃馆
烧烤专线 16,17,18,19 (淄博新区)
淄博新区
水晶街
原山快速路
百年课本博物馆
稷下湾文创街
黑铁山
九顶山
"六桥连七山"生态廊桥
花山
柳毅山
四宝山
牧龙山
玉皇山
淄博东收费站
玉黛湖生态乡村庄园
金岭回族镇
临淄收费站
临淄城市候机楼
临淄长途汽车站
临淄城区
九宫阁齐国文字博物馆
临淄大院鑫和博物馆
姜太公祠
管仲纪念馆
红螺山
蛟山
牛山
田齐王陵（四王
马莲台
鲁中城市候机楼
文昌古陶瓷博物馆
淄博市博物馆
八大局市场
张店城区
文旅专线
烧烤专线 11 (临淄)
淄博新区收费站
昌国快速路
淄博客运中心
烧烤专线 4,5,6,7,8,9,10 (中心城区)
淄博站
Zibo Railway Station
烧烤专线 20,21 (淄川)
Linzi District
炒米山生态风景区
张店东收费站
淄博市中心城区地图
见本图左下角插图
Please refer to lower-left Inset Map of Zibo Central Area
Zhangdian District
文昌湖收费站
金山
卧虎山
金山收费站
孝妇河
贾村水库
梓橦山景区
淄砚斋艺术馆
沾临高速淄川服务区
天堂寨景区
大北山
淄川城市候机楼
淄博市源一刻瓷艺术博物馆
淄川城区
北沈遗址
淄川博物馆
寨里窑址
淄川东收费站
长途汽车站
淄博煤矿博物馆
蒲松龄故居纪念馆
淄博矿务局德日建筑群
聊斋城景区
齐鲁欢乐世界
留仙谷景区
领尚琉璃文化创意园
洄村古楼
久润富硒生态科技园
聚相山景区
淄川龙泉收费站
玉皇山
济潍高速 S6
太河收费站
淄
川
区
Zichuan District
天明寺旅游风景区
大奎山
楼岱海
Haidai Pavilion
阁书钟
Zhongshuge Bookstore
"中国最美书店"钟书阁
The Most Beautiful Bookstore in China
西河收费站
齐山风景区
潭溪山旅游区
太河水库
马鞍山景区

博山东收费站
焦裕禄纪念馆
梦泉风景区
山 区
开元溶洞
坊
市
Weifa
(Pref
游度假区
鲁山收费站
八宝灵山寺
云水瑶风景区
沾临高速鲁山服务区
二郎山
三水源生态旅游度假区
鲁山国家森林公园
鲁山
1108
沂源鲁山溶洞群
风景区
沂源猿人遗址
神农药谷
锦水崖景区
青龙湖景区
六一八战备电台
旧址风景区
青兰高速 G22
沂源北收费站
沂源长途汽车站
沂源博物馆
沂源城市候机楼
沂源
县城
鲁村收费站
沂源东收费站
青兰高速
沂源服务区
沂源收费站
青兰高速沂源服务区
田庄水库
沂源县
Yiyuan County
汇泉桃花岛
牛郎织女景区
朱家户乡村旅游休闲区
中庄收费站
沾临高速中庄服务区
沂源苹果谷
沐心·双马山旅游度假区
红旗
水库
朱彦夫事迹
党性教育基地
鲁中桃花源
风景区

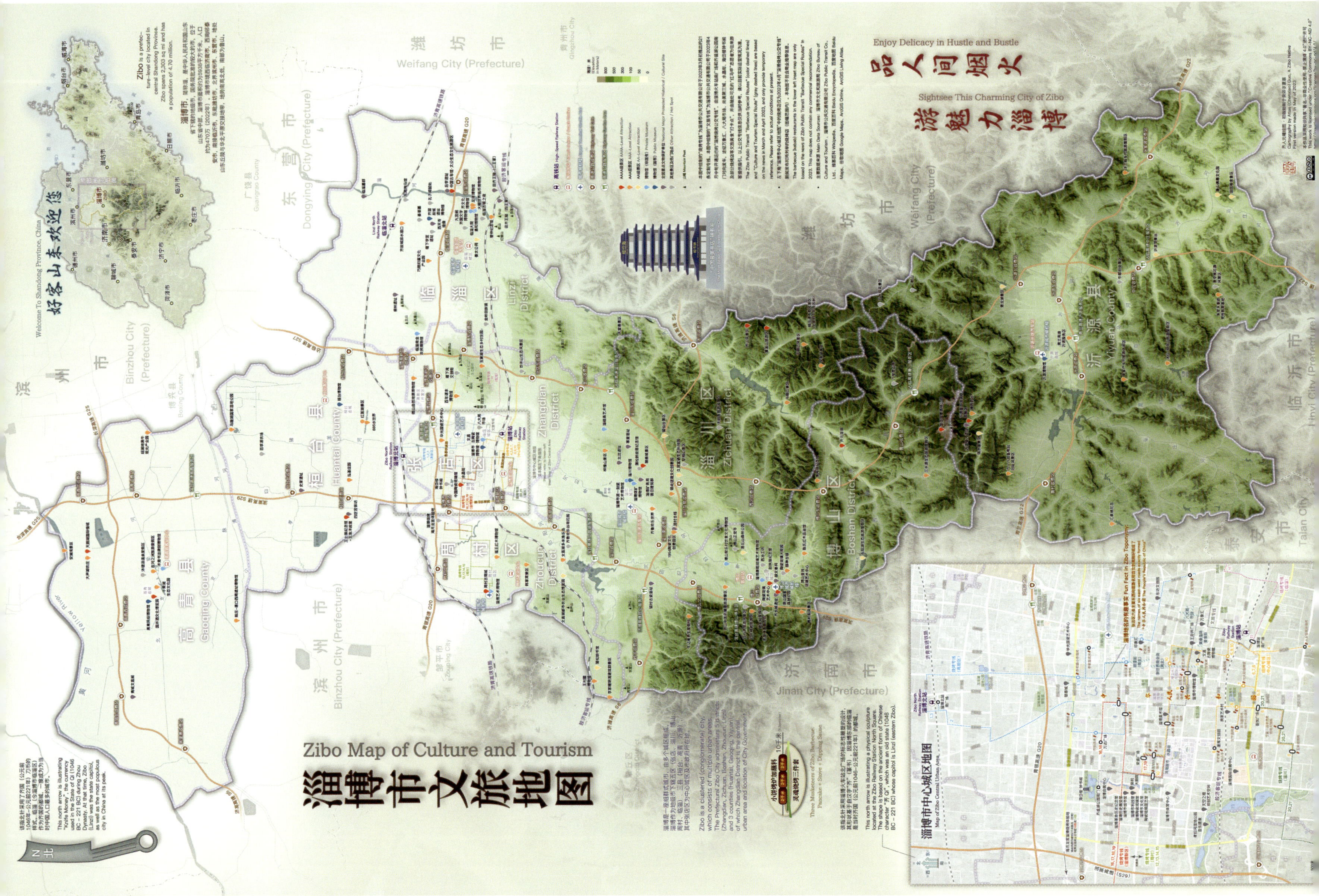
淄博市文旅地图
Zibo Map of Culture and Tourism
品人间烟火
Enjoy Delicacy in Hustle and Bustle
游魅力淄博
Sightsee This Charming City of Zibo
好客山东欢迎您
Welcome To Shandong Province, China
Zibo is a prefecture-level city located in central Shandong Province. Zibo spans 2,303 sq mi and has a population of 4.70 million.
淄博市，简称淄，是中华人民共和国山东省下辖的地级市，国务院批准的较大的市，位于山东省中部。淄博市面积约为5938平方千米，人口约为470万（2022年）。淄博市境西临济南市，西南邻泰安市，南接临沂市，东毗潍坊市，北界滨州市、东营市。地处山东丘陵与华北平原交接地带，地势南高北低，南部为鲁山。
淄博是一座组群式城市，由多个城区组成。淄博作为地级市下辖五区（张店、淄川、博山、周村、临淄）、三县（桓台、高青、沂源），其中张店区为中心城区及市政府所在地。
Zibo is a clustered (conglomerate) city, which consists of multiple urban areas. The Prefectural Zibo City administers 5 districts (Zhangdian, Zichuan, Boshan, Zhoucun, Linzi) and 3 counties (Huantai, Gaoqing, Yiyuan), of which Zhangdian District is the central urban area and location of City Government.
小饼烤炉加蘸料
灵魂烧烤三件套
Three Musketeers of Zibo Barbecue:
Pancake + Stove + Dipping Sause
10千米
该指北针采用淄博火车站北广场的标志性雕塑的设计，其形状基于自汉字"齐"（篆书），因淄博东部的临淄是当时齐国（公元前1046–公元前221年）的都城。
This north arrow is illustrating a physical sculpture located at the Zibo Railway Station North Square. The shape is based on the ancient form of Chinese character "齐 Qi", which was an old state (1046 BC – 221 BC) whose capitol is Linzi (eastern Zibo).
该指北针采用了齐国（公元前1046年–公元前221年）刀币的样式。临淄（今淄博市临淄区）作为齐国的都城，一度成为为当时中国人口最多的城市。
This north arrow is illustrating "Knife Money", the currency used in the State of Qi (1046 BC – 221 BC) during Zhou Dynasty. At that time, Zibo (Linzi) was the state capitol, as well as the most populous city in China at its peak.
N 北
临淄区
Linzi District
张店区
Zhangdian District
淄川区
Zichuan District
博山区
Boshan District
周村区
Zhoucun District
桓台县
Huantai County
高青县
Gaoqing County
沂源县
Yiyuan County
潍坊市
Weifang City (Prefecture)
东营市
Dongying City (Prefecture)
滨州市
Binzhou City (Prefecture)
济南市
Jinan City (Prefecture)
临沂市
Linyi City (Prefecture)
泰安市
Taian City
Yellow River
淄博市中心城区地图
Map of Zibo Central Urban Area
Zibo North Railway Station
淄博北站

《 Zibo Map of Culture and Tourism

Atlas (Chenxiao) Guo

My hometown, the city of Zibo in China, became a nationwide hotspot in the spring of 2023 due to its distinctive barbecue cuisine and honest, simple folk customs. I designed this map to introduce the history and culture of this beautiful city, showcasing its natural features, tourist attractions, public transportation, and fun facts. The initial version was created in Chinese; in subsequent revisions, some information was translated into English to serve international readers.

This is a reference map enriched with cultural and tourism information. The topographic base map combines shaded relief and stratified colors, presenting the natural landscape's beauty. It clearly illustrates the topographic characteristics while setting an elegant visual tone for terrain representation, with Zibo's major peaks and some suburban hills labeled.

In terms of symbol design, important tourism and transportation information is prioritized. This includes transportation hubs (railway stations, transit centers, airport shuttle hubs, highway toll gates, and service areas); transportation lines (high-speed railway and highway networks); and city bus "Barbecue Lines" and "Tourism Lines" specially designed in metro transit style. The map categorizes tourist attractions into different levels and types, such as museums, historic sites, and other popular spots.

The map includes two insets: a locator map of Shandong Province, showing administrative boundaries and physical geography; and a map of downtown Zibo, marking the main roads, business districts, featured barbecue restaurants, and a fun fact in local toponymy.

This map subtly integrates historical and cultural elements into its graphic design, such as barbecue kebabs combined with the map scales, and ancient knife-shaped money and a symbolic sculpture from railway station plaza as the north arrows. These reveal Zibo's rich history as the capital city of Qi State (1046–221 BCE), one of the most populated Chinese cities at that time. Landmark Haidai Tower, featuring Zhongshuge, known as the most beautiful bookstore in China, is specially showcased.

For me, as a native of Zibo, this map holds great significance, as it is tied to countless personal memories. With deep love and passion, I respectfully present my map as a gift to this lovely land, Zibo.

VOTELAND
VOTE YELLOW
VOTE PURPLE
WAFFLES
THE POST
The Voteland Post
ELECTION DAY

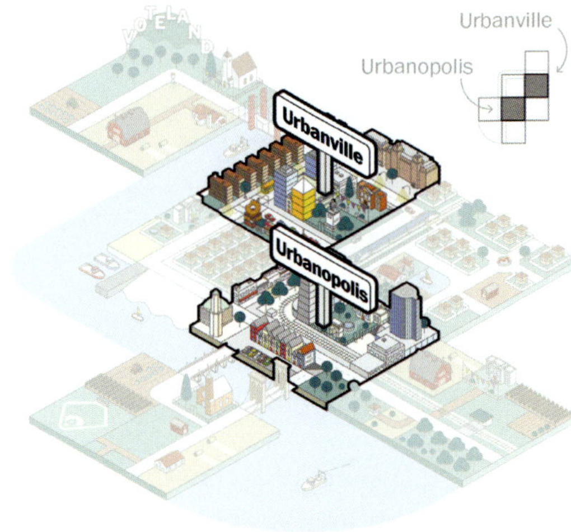
Urbanville
Urbanopolis

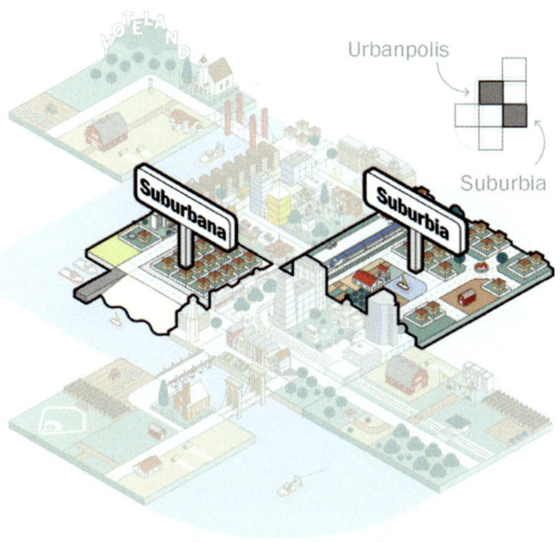

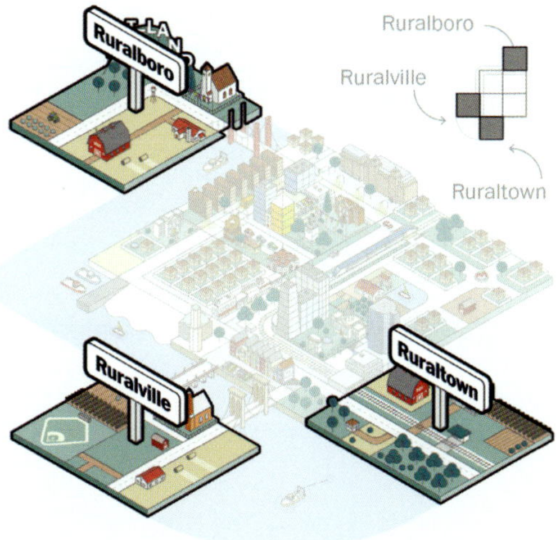

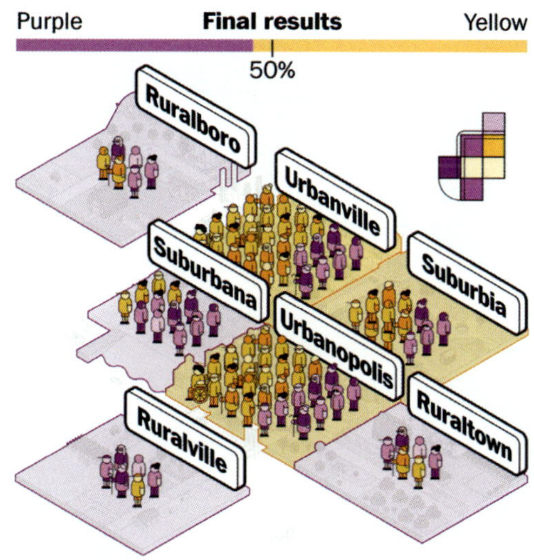

« Voteland

Adrián Blanco Ramos and Artur Galocha, The Washington Post

"Voteland" is an illustrated map that represents an imaginary country, used as the driver to explain a fairly complex technique: election modeling. The map is from a *Washington Post* article, "How Election Modeling Can Help Us Understand Who Might Win," originally published on May 24, 2022.

Election models, widely used in newspapers, help close the gap between the vote count at a moment in time and the final result, based on trends they spot in the available data. Yet, how election modeling works is an abstract concept that is hard to explain.

Using the "Voteland" metaphor of an imaginary country, readers can extrapolate how an election model works in any kind of election. With this map, we welcome the reader to Voteland, a fictional state heading to the polls to pick its new representative. Votelanders will decide between candidates from the Purple party and the Yellow party.

To bring the metaphor to life, we created different iterations of the map to show how Votelanders are spread unevenly across seven distinct counties. The detailed illustrations on an isometric and minimalist grid of seven imaginary regions helped us communicate how election modeling works during an election night. Each region contains different illustrated features that convey if the area is an urban or a rural region, as well as a few "Easter eggs" in the shape of urban elements from our own hometowns. In addition to other charts, we also designed a cartogram showing the final result of this fictional election and bringing closure to this explanatory piece.

An appealing and illustrated map can simplify a complex reality. Voteland's map graphics help to explain how mathematical modeling works in elections by making this information more digestible and easy to understand for readers. They are an excellent example of how we can distill complex concepts through good map design.

ADIRONDACK PAINT MAP
L. CHAMPLAIN
PLATTSBURGH
KEESEVILLE
DANNEMORA
MALONE
ST REGIS
CANTON, POTSDAM
UPPER CHATEAUGAY LAKE
AUSABLE FORKS
WILMINGTON
WHITEFACE MT.
CATAMOUNT MT.
BLOOMINGDALE
SARANAC LAKE
LAKE PLACID
PAUL SMITHS COLLEGE
SEE DETAIL
ST REGIS MT.
BLUE MOUNTAIN RD
ST. REGIS R.
RAINBOW L.
SARANAC VILLAGE (SEE DETAIL)
TUPPER LAKE
RAQUETTE R.
LONG L.
HUDSON R.
0 5 MILES
0 5 10 Km
GABRIELS
HOBART RD.
JONES P.
OSGOOD POND
PAUL SMITHS COLLEGE
VISITOR INTERPRETIVE CENTER
BLACK POND
KEESE MILLS ROAD
LOWER ST. REGIS L.
SPITFIRE L.
UPPER ST. REGIS LAKE
LAKE CLEAR
ADIRONDACK AIRPORT
NY RT 86
NY RT 30
NY RT 186
CO RT 30
CO RT 31
0 1 MILE
0 1 KM
SARANAC LAKE
SARANAC VILLAGE
NY RT 30
0 400 FEET
0 100 METERS

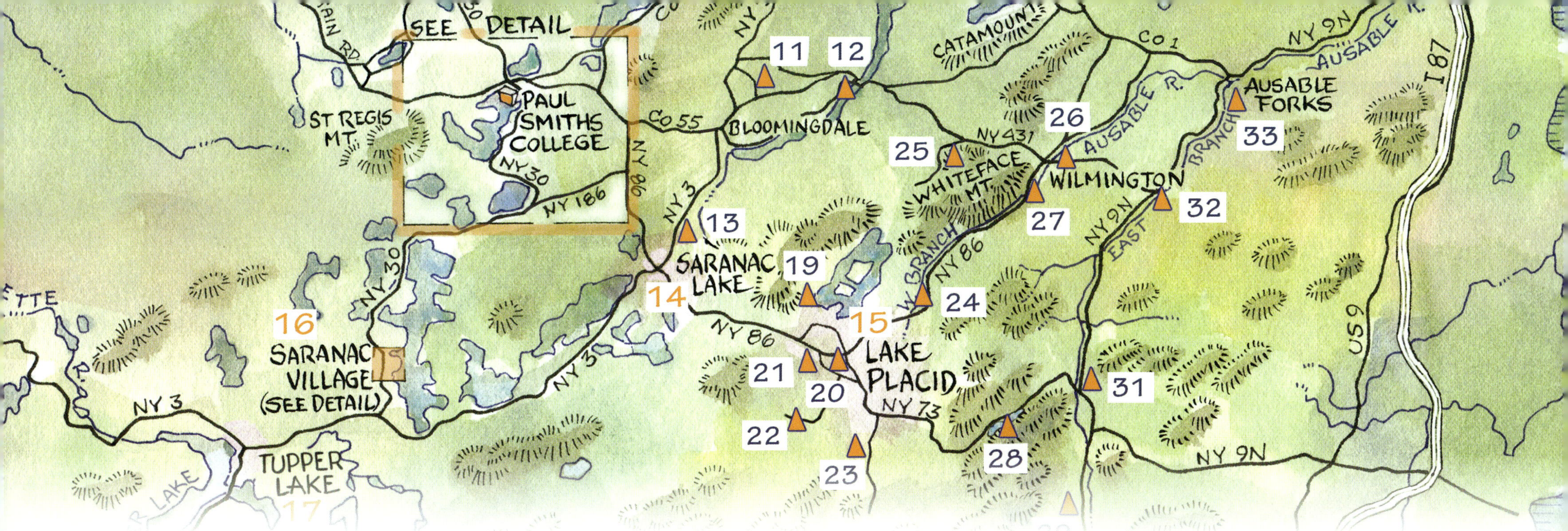

« Adirondack Paint Map

Jeff Mathison

"Adirondack Paint Map" grew out of frustration.

In 2021, I was attending the Publisher's Invitational Paintout for the second time. The event has been graciously hosted by Eric Rhodes, publisher of *Plein Air and Fine Art Connoisseur* magazines, for more than a decade. Up to a hundred artists convene at Paul Smith's College deep in the Adirondack forests of New York. We were there to paint the spectacular landscape, capturing the light as best as we could. No contest, no judgment, just a fellowship of artists.

However, the piney woods are traversed by twisty, hilly roads, and our only guide was a set of bland maps from Google. When I suggested to Eric that I use my mapping experience to make something better, he agreed it was worth a try.

Online free topographic mapping gave me a start. I decided on a scale, assembled dozens of screen-shots in Photoshop, and printed a base map of landforms and roadways to trace. I didn't try to get the elevations accurate, I wanted to give an arty feel—like "here be mountains." I did aim for reasonable accuracy with the roads. A driver, new to the area, could at least see a hard curve on the map translated into reality, and feel assured of the right path. Likewise, an accurate scale is critical for planning the day's painting adventure. Most of all, an overall map gives people a sense of the whole area, not just what they see through their windshield as GPS tells them where to turn.

Space was very limited, so I used number labels corresponding to text in the guidebook. For fun, I added pictures of the wildlife inhabitants and a North arrow befitting the region!

The map has been in use for the Publisher's Invitational events in 2023 and 2024 as well as the Fall Colors gathering in the same area. The Google maps are gone from the guidebook. The key now includes photos of the locations and, of course, addresses that can be plugged into GPS. I have not been able to attend recently, but I've heard from a number of artists that the map is "awesome" and very helpful!

Gunung Mulu National Park »

Phoebe Ly

I am proud to present my first map designed for print, showcasing Taman Negara Gunung Mulu (Gunung Mulu National Park), a UNESCO World Heritage Site in Sarawak, Malaysia—my home country. While many exquisite maps focus on parks in North America and Europe, there are few maps of parks in Malaysia or even Southeast Asia. This map aims to educate readers and to inspire a deeper appreciation for Taman Negara Gunung Mulu's rich biodiversity and spectacular cave systems, through a well-designed and easily comprehensible poster.

Taman Negara Gunung Mulu is a remote wilderness with diverse wildlife and physically challenging landscapes. The map features an oblique, panoramic view that captures the park's mountainous terrain and allows viewers to appreciate the park's vastness. The design theme evokes a lush tropical forest, with a carefully selected color gradient to reflect the vibrant topography. Elements of Indigenous Sarawakian tribal aesthetics are incorporated through the map's border, typography, and colors, adding a layer of personality to the design.

Due to limited Malaysian open data resources, I utilized information from OpenStreetMap, All Trails, and the Mulu Caves Project to map major trails, rivers, caves, and mountains.

Although the map includes fewer geographic features, it is enriched with informative text and custom illustrations highlighting the park's biodiversity, landforms, and geology. The park's remarkable cave system is illustrated in a papercut style, offering depth while maintaining simplicity. Balancing text and illustrations was crucial to ensure that the information would remain readable and engaging. I varied the illustration styles and detail levels to make certain elements stand out while others blend seamlessly into the background. The illustration of bats was actually something I knew I wanted from the beginning to draw attention, while the section on landforms was one of the last parts added because, although I was struggling with limited space, it was important to me to include the landforms as part of the park's diversity. I wanted to show this in a minimalistic and consistent way, similar to icons, so I displayed the outlines of each landform within a square. Displaying them together vertically was initially due to space, but had the added benefit of looking like a cool stamp collection.

BIODIVERSITY
NUMBER OF SPECIES
FAUNA
20,000 Invertebrates
76 Amphibians
270 Birds
55 Reptiles
81 Mammals
48 Fish
FLORA
3,500 Vascular Plants
1,700 Mosses, Liverworts
1,500 Flowering Plants
8,000 Fungi
442 Pteridophytes
109 Palms
8 species of hornbill have been spotted in the park. This is a Rhinoceros hornbill, the official state bird of Sarawak!
TAMAN NEGARA
GUNUNG MULU
UNESCO WORLD HERITAGE SITE
Sarawak, Malaysia
Taman Negara Gunung Mulu, situated in the Malaysian state of Sarawak on the island of Borneo, provides protection for a substantial area of Borneo's primary tropical forest and a home for a high diversity of species, including many endemics and threatened species, the large cave passages and chambers provide a major wildlife spectacle in terms of millions of cave swiftlets and bats. It is outstanding both for its high biodiversity and for its karst features. The park contains seventeen vegetation zones and its palm species are exceptionally rich, making it one of the worlds richest sites for palm species.
Established: 1974
Land Area: 52,864 hectares
Highest Elevation: 2,376 metres
Average Yearly Rainfall: 3,839 millimeters
Lowland Temperature Range: 23°C to 26°C
Summit Temperature Range: 14°C to 18°C
0 2 4 KM
Data Sources: Cave maps - Mulu Caves Project; DEM - NASA SRTM V003; Boundaries, Rivers, Places - ©OpenStreetMapcontributors. Trails - AllTrails; Geology - Peta Geologi Sarawak 2013 by jmg.gov. Text references: Gunung Mulu National Park. (1 May 2023). in Wikipedia (CC-BY-SA 4.0); WHC UNESCO. (30 Nov 2000) (CC-BY-SA IGO 3.0). https://whc.unesco.org/en/list/1013. Images: Hornbill and bats - Freepik; Landforms adapted from LittleRed on TPT; Border pattern - Uldownload.
Created by Phoebe Ly, June 2023
Gunung Api is notable for its pitcher plant diversity with at least 8 different species
GUNUNG BENARAT 1,585 M
GUNUNG API 1,750 M
GUNUNG MULU 2,376 M
BRUNEI
Colony of over 3 million free tailed bats (Chaerephon plicata)
The park is only accessible by air through Mulu Airport
PARK HEADQUARTER
HEADHUNTER'S TRAIL
PINNACLES TRAIL
MULU TRAIL
SUNGAI MELINAU
SUNGAI MELINAU PAKU
SUNGAI TUTOH
PROPOSED PARK EXTENSION
The park is dominated by Gunung Mulu, the second highest mountain in Sarawak, a 2,376 m high sandstone pinnacle. It is the most studied tropical karst area in the world. The geological Melinau Formation contains a remarkable concentration of caves, revealing a geological history of over more than 1.5 million years.
The Sarawak Chamber, 600 m by 415 m and 80 m high, is the largest known cave chamber in the world. Deer Cave is the second largest cave passage in the world with the largest number bat species in a cave. It is home to the largest colony of cave swivelets.
LANDFORMS
Karst Towers
Rugged Summits
Caves
Waterfalls
Gorges
Sheer Cliffs
Floodplains
GEOLOGY
PzMl Melinau Limestone
Mulu Formation (Sandstone & shale)
PzS Setap Shale
NzBe Belait Formation (Mainly sandstone)
Terrace Gravels
CAVE SYSTEM
Since 1978, this park has 295 km of explored caves. The caves are still being explored and most of this is done through the Mulu Caves Project.
3 notable caves
1 Sarawak Chamber
12 million cubic metres
80 m High
600 m Long
Cave chamber inside Good Luck Cave
2 Deer Cave
146 m High
Main Passage Entrance
3 Clearwater Cave
236 km Length cave system (9th longest in the world)
MALAYSIA

LOCURILE APUSE
ale Bucureștiului
The Forgotten Places of Bucharest
LOCURI APUSE
FORGOTTEN PLACES
Zone verzi, plantații, grădini care înconjurau casele unor personalități ori grădini de vară, sau chiar curți domnești, au început să apară încă din 1400 (Grădina Domnească), dar au dispărut cu preponderență în secolele XIX și XX.
Green areas, plantations, gardens that surrounded the houses of some personalities or summer gardens, or even princely courts, began to appear as early as 1499 (Grădina Domnească), but mostly disappeared in the 19th and 20th centuries.
GRĂDINI
GARDENS
FABRICI
FACTORIES
Multe dintre fabricile din București își au începuturile în sec. XVIII și au dispărut odată cu dezindustrializarea.
Many of the factories in Bucharest have their beginnings in the 18th century and disappeared with deindustrialization.
Gările din oraș au început să apară în 1869, dar treptat unele au dispărut ori și-au pierdut funcționalitatea. Principalele gări au fost amplasate în funcție de punctele cardinale, urmând modelul francez.
Train stations in the city started to appear in 1869 and gradually some disappeared or lost their functionality. The main railway stations were located according to the cardinal points, following the French model.
GĂRI
RAILWAY STATIONS
LEGENDĂ | LEGEND
Loc actual, cu aspect/denumire/funcționalitate schimbată
Current place, with changed appearance/name/functionality
Loc actual: recomandat a fi văzut/vizitat
Current place: recommended to be seen/visited
GARĂ FUNCȚIONALĂ
Functional train station
Clădire dispărută/demolată
Disappeared/Demolished building
Biserică demolată
Demolished church
FOSTĂ GARĂ
Former train station
Grădină dispărută
Disappeared garden
HANURI
INNS
Hanurile bucureștene au apărut începând cu secolul XVII, erau locuri primitoare pentru străini sau călugări, având și un rol important în comerț, care s-a pierdut odată cu a doua jumătate a secolului XIX.
Bucharest inns appeared in the 17th century, they were welcoming places for foreigners or monks, also having an important role in trade, which was lost with the second half of the 19th century.
SURSE DE DATE & INSPIRAȚIE | DATA SOURCES & INSPIRATION:
Open Street Map, 2022
Lista Monumentelor Istorice, 2015
Planul Bucureștiului, Maior Baron Rudolf Borroczyn, 1846
Planul Bucureștiului, Maior Dimitrie Papazoglu, 1871
Planul Orașului București, Serviciul Geografic al Armatei, 1911
Istoria Bucureștilor din cele mai vechi timpuri pînă în zilele noastre, Constantin C. Giurescu, Editura pentru Literatură, București, 1966
* Ilustrațiile reprezintă clădiri emblematice care există și astăzi și amintesc de trecut.
* Datele nu sunt complete, există și alte locuri care au dispărut sau au fost uitate.
* Unele dintre locurile actuale au fost recomandate spre a fi vizitate fiindcă sunt mai puțin cunoscute.
Hartă creată de Oana Candit, Esri România, 2023
Aproximativ 20 de lăcașuri de cult au fost distruse complet în comunism, în anii '80, din dorința de sistematizare.
About 20 places of worship were completely destroyed during communism, in the 80s, due to the desire for systematization.
BISERICI
CHURCH
Etichete cheie
Key labels
1. Palatul Cotroceni
2. Azilul Elena Doamna
3. Fabrica de Ciocolată Frank Kafe
4. Fabrica de Ciocolată Anghelescu
5. Pavilionul Seismologic
6. Sanatoria de Tuberculoși
7. Biserica Albă Postăvari
8. Palatul Crețulescu
9. Casa Grădișteanu-Ghica
10. Casa Filipescu-Cesianu
11. Muzeul Național de Istorie Naturală
12. Muzeul Național al Țăranului Român
13. Muzeul Geologic Național
14. Vila Filipescu-Brâncoveanu
15. Muzeul de Artă Zambaccian
16. Muzeul Memorial Dr. Victor Babeș
17. Muzeul Național al Hărților și Cărții Vechi
18. Fabrica de Frânghii
19. Moara Stoncovici
20. Observatorul Astronomic Amiral Vasile Urseanu
21. Casa Grigore Petrescu
22. Muzeul Simu
23. Ateneul Român
24. Biserica Enei (Ienii)
25. Mănăstirea Sfântul Sava
26. Biserica Sărindar
27. Palatul Cercului Militar Național
28. Palatul Eforiei Spitalelor Civile
29. Hanul Serafim
30. Palatul Băncii Marmorosch Blank
31. Hanul Filipescu
32. Hanul Constantin-Vodă
33. Hanul Stavropoleos
34. Hanul Șerban-Vodă
35. Hanul Colței
36. Hanul cu Tei
37. Hanul Gabroveni
38. Hanul Papazoglu
39. Mănăstirea Stelea
40. Templul Coral
41. Biserica Sf. Vineri
42. Hanul lui Manuc
43. Biserica Sf. Nicolae Sârbi
44. Biserica Sf. Nicolae Jitnița
45. Biserica Bradu Staicu
300 m

« The Forgotten Places of Bucharest

Oana Candit, Esri Romania

Bucharest, the capital city of Romania, used to be known for its beautiful gardens, which were marked on old maps and plans. The managing director at Esri Romania, Cristian Vasile, found inspiration in a plan from 1846, created by Rudolf Artur Borroczyn, that illustrated in great detail the gardens and vineyards of the city. Motivated by this, I wanted to create a new cartographic style and product that brings together places from different periods.

Many of the locations on the map could best be described by the word "sunset," which in Romanian has a figurative meaning of "disappeared" or "passed." This word drove my map design decisions, as I wanted to create an experience for discovering the old soul of the city. The gold-colored and blurred edges of the symbols are the map's most distinctive features. The transparent map borders enhance the idea that we see through some fogged memories, and the round shapes remind us of the sun.

The disappeared gardens reflect the sun's colors, and their edges are blurred to spread the colors because their limits could not be precise. The background color of the map reminds readers of other gardens and vineyards that could not be displayed but are a part of the city's history.

Similar glow effects in yellow can be noticed on the map for symbols, representing churches, railway stations, or other buildings. The pins for vanished or forgotten places, unlike those for current places recommended to be visited, are not filled with color, because I wanted to highlight that something is missing. Inspired by the 1871 plan of Bucharest by Major D. Pappasoglu, I drew illustrations for some iconic buildings that still stand today and are more or less well known. The base map contains current data, including green spaces, streets, and subway entrances, making it easier for users to accurately locate disappeared places when using the map in the field.

The yellow accent colors and fading graphics or symbols in the background create a striking contrast. These places yearn to illuminate our memories or feelings in a new way, and this map helps prevent them from fading into obscurity once again. It serves as a reminder of all the places we cherish.

AOTEAROA
NEW ZEALAND
Cities, towns, and localities with significant land and bathymetric features
N
TE IKA-A-MĀUI /
NORTH ISLAND
TASMAN SEA
TE WAIPOUNAMU /
SOUTH ISLAND
RAKIURA /
STEWART ISLAND
SOUTH PACIFIC
OCEAN

40° S
42° S
44° S
Cook Cha
Waiatoto Channel
Abut
Okarito
Gillespies
Point/Kōhaihai
Makawhio Point
Heretaniwha Point
Jackson Head
Okahu
Cascade Point
Barn Knob
Awarua Point
Big Bay
Martins Bay
Yates Point
Milford Sound
Saint Anne Point
Wanaka
Queenstown
Cromwell
Omakau
Clyde
Alexandra
Te Anau
Manapouri
Mossburn
The Key
Secretary I.
Round Head
Roxburgh
Ettrick
Clarks Junction
Patearoa

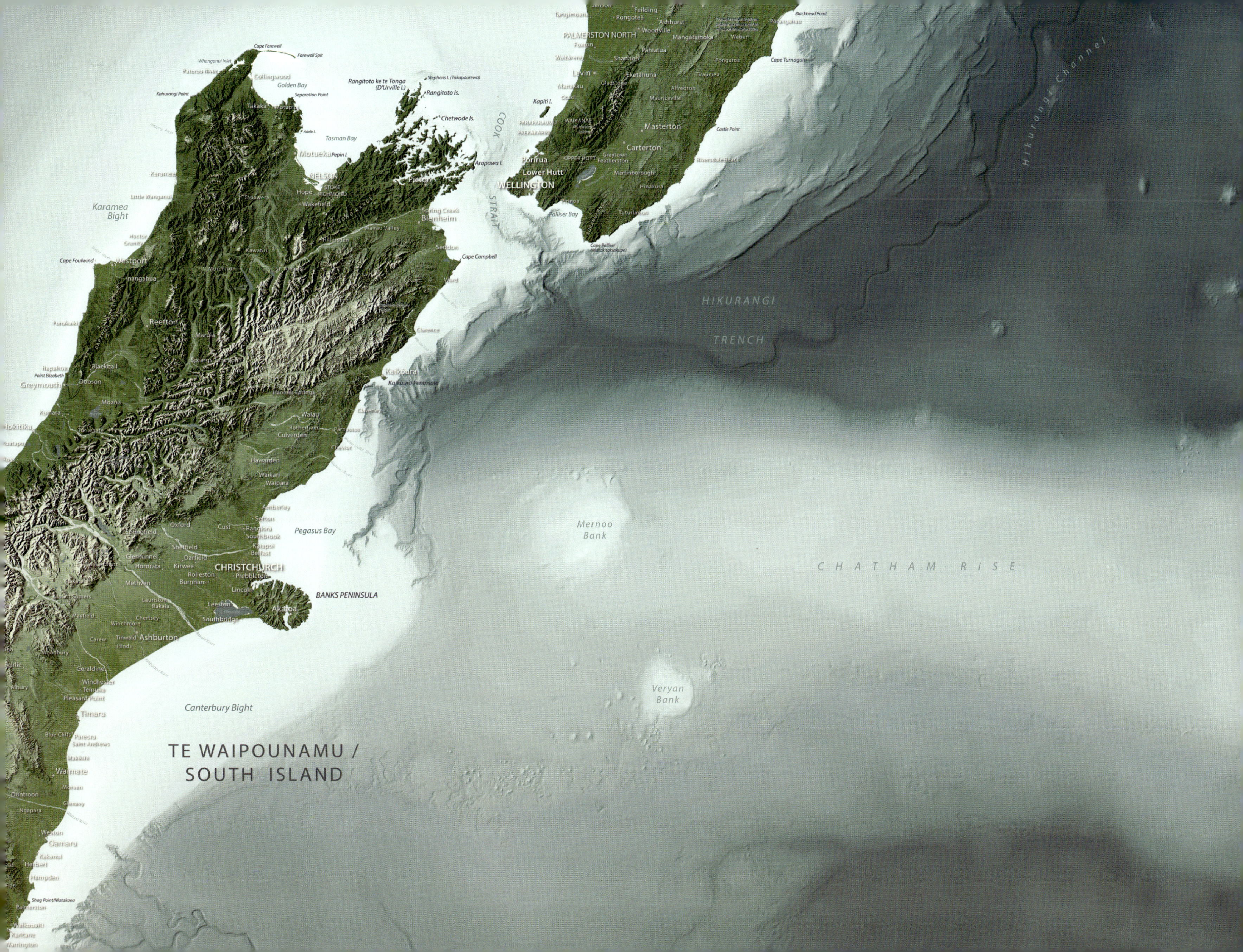
TE WAIPOUNAMU /
SOUTH ISLAND
PALMERSTON NORTH
Levin
Masterton
Carterton
Porirua
Lower Hutt
WELLINGTON
COOK
STRAIT
Cape Farewell
Farewell Spit
Golden Bay
Tasman Bay
Rangitoto ke te Tonga
(D'Urville I.)
Rangitoto Is.
Stephens I. (Takapourewa)
Chetwode Is.
Arapawa I.
Kapiti I.
Cape Turnagain
Castle Point
Palliser Bay
Karamea
Bight
Motueka
NELSON
Blenheim
Cape Campbell
Westport
Cape Foulwind
Reefton
Greymouth
Hokitika
Kaikōura
Kaikōura Peninsula
Clarence
Culverden
Hawarden
Amberley
Pegasus Bay
CHRISTCHURCH
BANKS PENINSULA
Akaroa
Ashburton
Geraldine
Timaru
Canterbury Bight
Waimate
Oamaru
Shag Point/Matakaea
HIKURANGI
TRENCH
Hikurangi Channel
Mernoo
Bank
Veryan
Bank
C H A T H A M   R I S E

Aotearoa / New Zealand »

Ian Reese, XYCarto

New Zealand is a nation graced by dramatic landscapes, but subject to the oceans that surround it. The intention of this map is to celebrate the terrestrial and begin to incorporate a new story of the waters beyond the shores.

For a number of years, I explored different iterations of this map whenever I had time available. I played with deep contrasts for the bathymetry, multiple color palettes, and various methods of layering data. My overall goal always remained the same, however: how do I highlight features without overemphasizing scale?

This is achieved through careful contrasting, blending, and taking a "less is more" approach. This map contains multiple layers specifically designed to provide a contrast between transitions. For example, where the ocean meets the land, the bathymetry is layered several times incorporating different blending techniques. The result is a transition with a clear delineation between land and sea that also acts as a frame for the terrestrial story. As for the "less is more" approach, this is primarily achieved by first developing a very data-rich map, then systematically removing one layer at a time and assessing whether the map still achieves its message. Over time, I have found methods like significant reductions in forest layers that allow for more readability of the land features.

The map is a blend of the traditional topographic map style and a degree of artistic license. Labels follow common methods for hierarchical layout, elevations follow a palette similar to elevation maps, and data are derived from official topographic sources. With this as my base, I tend to drift to the edges of steadfast cartographic rules. The map should always remain a source of knowledge and information, maintaining the greatest accuracy possible; however, if I need to invert a color gradient to make a feature stand out, push my stops into an illogical pattern, or highlight regions by layering data, I do it. I want the land to speak; sometimes it just needs a little nudge.

-A-MĀUI /
H ISLAND
Kaipara Harbour
Warkworth
Kawau I.
Cape Colville
Hauraki Gulf
Great Mercury I. (Ahuahu)
Mercury Is. (Iles d'Haussez)
Red Mercury I. (Whakau)
Mercury Bay
COROMANDEL PENINSULA
AUCKLAND
Manukau
Manukau Harbour
Thames
Pukekohe
Mayor I. (Tuhua)
Waihi Beach
Huntly
Te Aroha
Matakana I.
Motiti I.
Whakaari/White I.
Bay of Plenty
Cape Runaway
HAMILTON
TAURANGA
Cambridge
Morrinsville
Matamata
Kawhia Harbour
Te Awamutu
Tokoroa
ROTORUA
Whakatāne
Te Kuiti
Edgecumbe
Taupō
Acacia Bay
Lake Taupō
Tūrangi
Gisborne
Te Karaka
North Taranaki Bight
Taumarunui
NEW PLYMOUTH
Cape Egmont
Stratford
Ōpunake
Hāwera
Raetihi
Ohakune
Taihape
MĀHIA PENINSULA
Wairoa
Hawke Bay
NAPIER
HASTINGS
Cape Kidnappers
South Taranaki Bight
WHANGANUI
Marton
Dannevirke
Waipukurau
PALMERSTON NORTH
Feilding
Levin
Kapiti I.
Masterton
Carterton
Porirua
Lower Hutt
WELLINGTON
COOK
STRAIT
Blenheim
Picton
Tasman Bay
Rangitoto ke te Tonga (D'Urville I.)
Cape Campbell
Cape Palliser
Cape Turnagain
Castle Point
Hikurangi

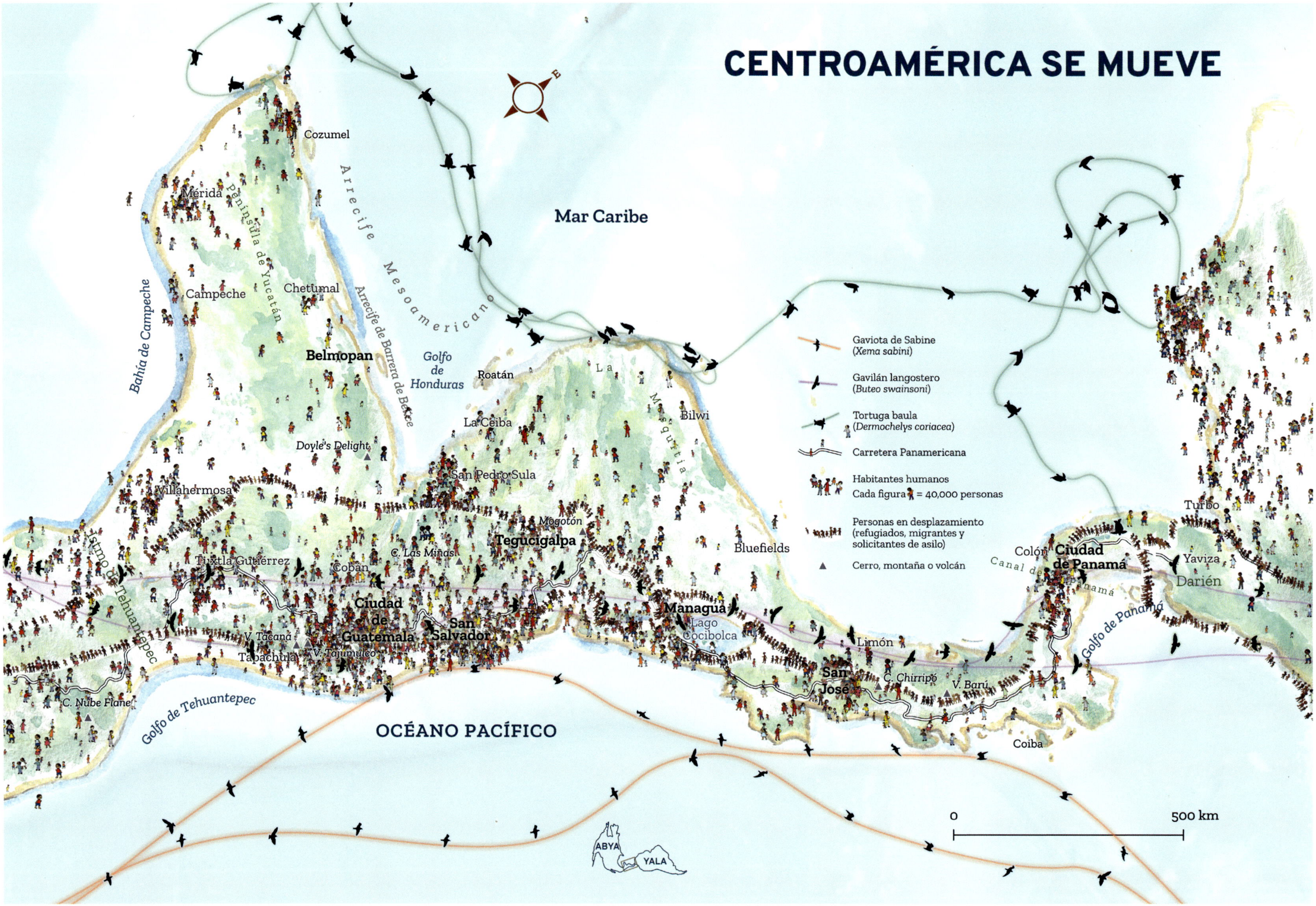
CENTROAMÉRICA SE MUEVE
Mar Caribe
OCÉANO PACÍFICO
Gaviota de Sabine
(Xema sabini)
Gavilán langostero
(Buteo swainsoni)
Tortuga baula
(Dermochelys coriacea)
Carretera Panamericana
Habitantes humanos
Cada figura = 40,000 personas
Personas en desplazamiento
(refugiados, migrantes y solicitantes de asilo)
Cerro, montaña o volcán
0
500 km
Cozumel
Mérida
Península de Yucatán
Campeche
Chetumal
Bahía de Campeche
Arrecife Mesoamericano
Arrecife de Barrera de Belice
Belmopan
Golfo de Honduras
Roatán
La Ceiba
Doyle's Delight
San Pedro Sula
La Mosquitia
Bilwi
Villahermosa
Istmo de Tehuantepec
Tuxtla Gutiérrez
Cobán
C. Las Minas
Mogotón
Tegucigalpa
Bluefields
Ciudad de Guatemala
San Salvador
Managua
Lago Cocibolca
V. Tacaná
V. Tajumulco
Tapachula
C. Nube Flane
Golfo de Tehuantepec
Limón
San José
C. Chirripó
V. Barú
Coiba
Colón
Ciudad de Panamá
Canal de Panamá
Golfo de Panamá
Yaviza
Darién
Turbo
ABYA
YALA

« Centroamérica se mueve (Central America Moves)

Gabriela Fenner Sánchez and Mir Rodríguez Lombardo, Colectiva Cartográfica de los Sures Globales

This map was commissioned by *Revista de la Universidad de México* for its July 2023 issue on Central America. The editors simply told us, "Make a map of Central America." We began by discussing which stories we wanted to tell. One topic that particularly interests us is the mobility of living beings, both human and nonhuman, across the isthmus. In the context of environmental and social crises, this territory has once again become a critical corridor.

We drew the most common path of human groups, which has a general south-to-north direction. Data were gathered from various sources; media reports were of great help, as there are very few official sources and routes change all the time. This flow is different from the round-trip animal routes, both by air and by sea. We picked one species per ocean, for visual balance, and one land route to show diversity of movement. Bird migration data are easy to find with Audubon's Bird Migration Explorer, but for other species one must dig into specific studies.

Although Central America is often referred to as a "bridge," for us it is also a highly diverse inhabited space. To show this dual dimension, we represented settled human population density with figures from the work of Salvadoran painter Maya Salarrué (1924–95). We used data from Gridded Population of the World and created skin-tone distributions with a bit of Python.

We wanted the map to be beautiful and to include the warmth of handmade work, which has been largely lost with geographic information systems (GIS). Yet, at the same time, we did not want to give up the capabilities of GIS for management, representation, and data analysis. Thus, the result is a hybrid of these two methodological and aesthetic approaches.

Finally, a point of interest for our collective is the questioning of the norms of conventional cartography, which is inherently colonial. The orientation of this map, for example, not only appeals to us aesthetically but also reduces the latitudinal hierarchies prevalent in maps of the continent of Abya Yala. We have also erased the national borders imposed on the territory, emphasizing geographical continuity.

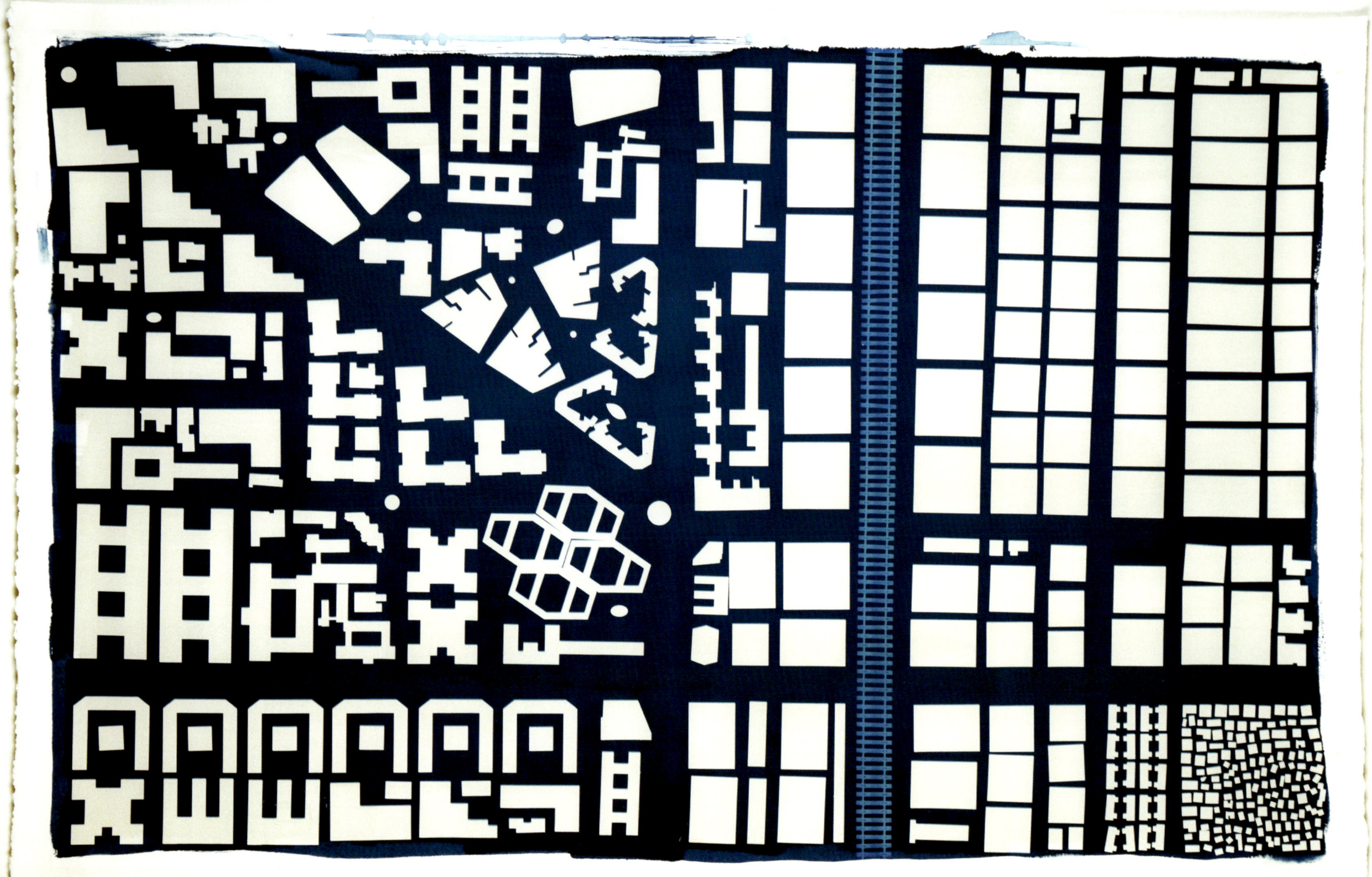

« Blueprint: City Grid

Svea Josephy, Michaelis School of Fine Art, University of Cape Town

The "Blueprint" project draws critical attention to the notion that the legacy of apartheid city planning is still strongly with us in South Africa today. One can clearly see the "buffer zones," green belts, railway lines, freeways, and industrial areas that were created to keep people apart. Considering South African cities has also helped me to understand some of the creative uses to which people have put the city. Spaces are being used inventively in ways that were never intended by the bureaucrats who initially produced them.

My "Blueprint" series is connected to the photographic work I have been making for the last fifteen years. While my other projects have involved going out into cities to photograph, "Blueprint" is about staying in one place and thinking about cities. I arrived at this "form fits function" moment, where the materiality of the cyanotype worked together with the concept of drawing apartheid-era planning into focus. So, working with aerial photographs, maps, documents from archives, and photographs, I began the series.

I draw the structures digitally by hand and then cut them out of mounting board with a laser cutter, leaving me with thousands of individual pieces. Each shape references a mansion, house, garage, block of flats, informal dwelling, shop, school, shopping mall, hospital, or clinic. In my work as a photographer, I am out in the field, engaging with the life of the city, the noise, the chaos. In contrast, the darkroom is a ritualistic and meditative space. It allows me to contemplate the city as I map it: How is it laid out? Why? Who lives where? What footprint do they occupy? What is the shape of each structure? Is it big or small? Where are the services and recreational spaces? I have many images around me as I lay down shapes in relation to each other in gridded or informal arrangements. Each cyanotype is original because it is constructed differently on the page, so it's unlike conventional photography, which can be infinitely reproduced. It holds the contradiction of the original blueprints in planning, architecture, and industry, which were used for their accurate reproducibility, and yet a blueprint was also considered to be something secret and precious.

A New and Accurate Map of Earth's Moon »

Kevin Sheehan, Manuscript Maps

I am too young to have witnessed the "space race," but I grew up fascinated by space science and exploration. The fanfare around the fiftieth anniversary of Apollo 11, and the excitement of the many upcoming missions to get humans back to the Moon, inspired me to draw this map, which is the largest and most detailed one I have yet created.

My dip-pen-and-ink cartographic style is intended to create up-to-date maps with the aesthetics of antique cartography. The first task was to decide on a projection, and because we experience the Moon in two halves—the near side and the far side—a double-hemisphere "planisphere" style made sense. In addition to natural features, I wanted to map all the missions to the Moon: robotic and human landings, plus flybys and orbits. Furthermore, I wanted to include as many diagrams and as much information as possible. The cosmic clockwork of the Moon, Earth, and Sun is complex, and I spent many hours researching these natural mechanics—libration, lunation, orbital periods, eclipses—and how to create diagrams that communicated them effectively, accurately, and beautifully. I must give thanks to NASA, National Geographic, and the USGS for the innumerable freely available photos and maps of the lunar surface, informative diagrams, and other information, without which it would have been far more difficult to create this map.

The aesthetics of my maps are as important to me as the content. I drew several plans to work out how to fit everything together into a single cohesive, harmonious, beautiful piece of cartography. Researching, planning, and drawing the final pencil-on-paper version took two months. Drawing in ink took a further four, including an entire month just stippling the craters of the lunar surface.

Great design creates a great map. As a work of art, a map should appear balanced and beautiful from a distance, beckoning the viewer to have a close look. As a work of cartography it should effortlessly communicate a wealth of accurate information, engaging the viewer to closely explore its content. I hope that with my "Map of Earth's Moon" I have accomplished both great art and great cartography.

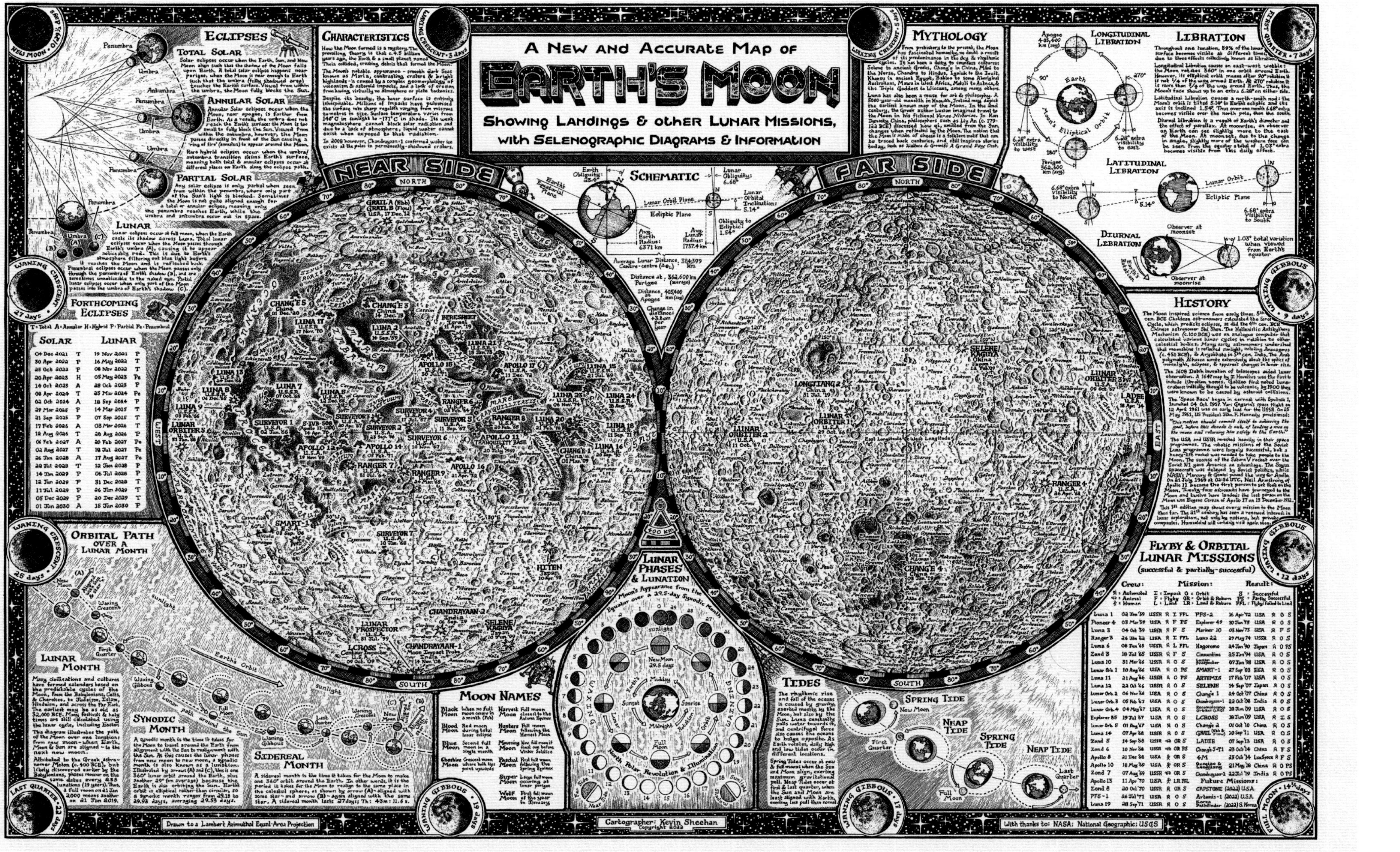
A New and Accurate Map of
Earth's Moon
Showing Landings & other Lunar Missions, with Selenographic Diagrams & Information
Eclipses
Total Solar
Annular Solar
Partial Solar
Lunar
Forthcoming Eclipses
Characteristics
Mythology
Longitudinal Libration
Libration
Latitudinal Libration
Diurnal Libration
Schematic
Near Side
Far Side
North
South
West
East
History
Flyby & Orbital Lunar Missions (successful & partially-successful)
Orbital Path over a Lunar Month
Lunar Month
Synodic Month
Sidereal Month
Moon Names
Lunar Phases & Lunation
Tides
Spring Tide
Neap Tide
New Moon
First Quarter
Full Moon
Last Quarter
Drawn to a Lambert Azimuthal Equal-Area Projection
Cartographer: Kevin Sheehan
With thanks to: NASA; National Geographic; USGS

meaning both total & annular eclipses occur at different places on Earth along the eclipse path.

PARTIAL SOLAR

Any solar eclipse is only partial when seen from within the penumbra, where only part of the Sun's light is blocked. Sometimes the Moon is not quite aligned enough for a total or annular eclipse, meaning only he penumbra reaches Earth, while the ımbra and antumbra occur out in space.

occur at full moon, when the Earth dow across Luna. Total lunar when the Moon passes through (A), causing it to appear This is due to Earth's g out blue light before d is reflected back. the Moon passes only shadow (B), and are naked eye. Partial part of the Moon 's shadow (C).

Penumbral

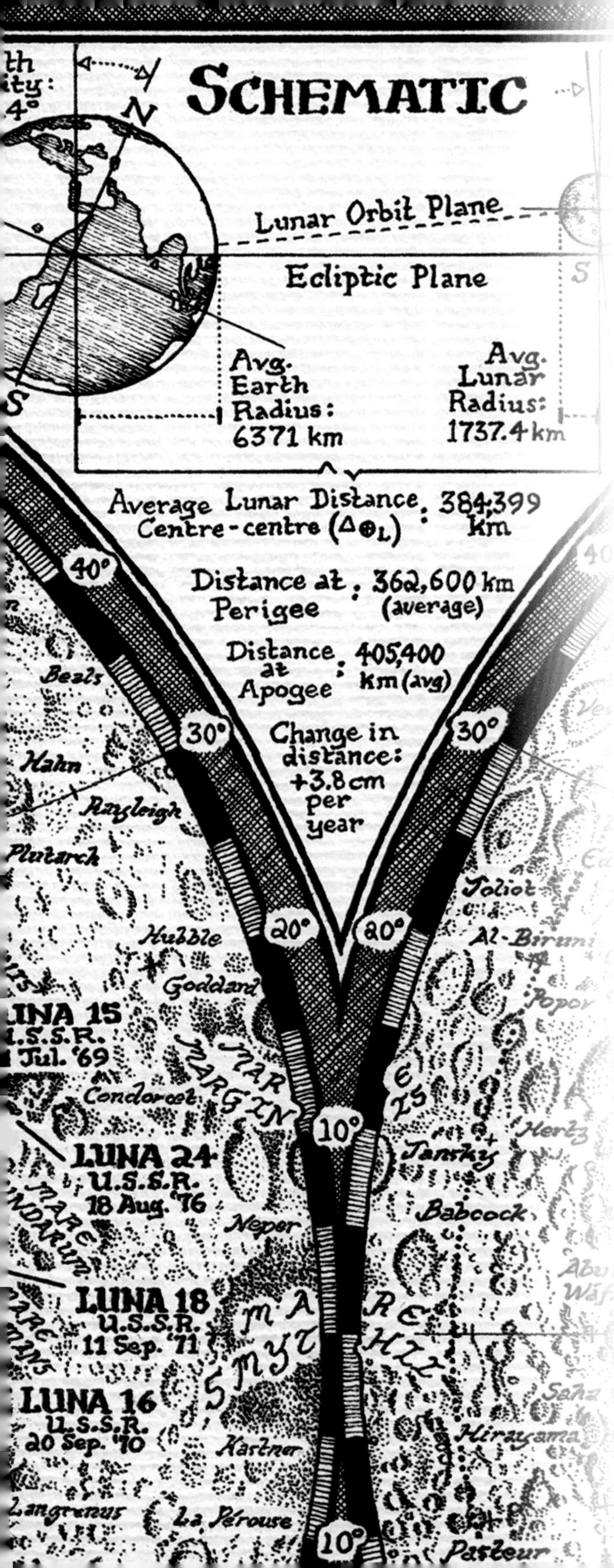
SCHEMATIC
N
S
Lunar Orbit Plane
Ecliptic Plane
Avg. Earth Radius: 6371 km
Avg. Lunar Radius: 1737.4 km
Average Lunar Distance Centre-centre: 384,399 km
Distance at Perigee: 362,600 km (average)
Distance at Apogee: 405,400 km (avg)
Change in distance: +3.8 cm per year
40°
30°
20°
10°
Beals
Hahn
Rayleigh
Plutarch
Hubble
Goddard
Joliot
Al-Biruni
LUNA 15 U.S.S.R. Jul. '69
Condorcet
MARE MARGINIS
LUNA 24 U.S.S.R. 18 Aug. '76
Neper
Jansky
Hertz
Babcock
LUNA 18 U.S.S.R. 11 Sep. '71
MARE SMYTHII
LUNA 16 U.S.S.R. 20 Sep. '70
Kästner
Hirayama
Langrenus
La Pérouse
Pasteur

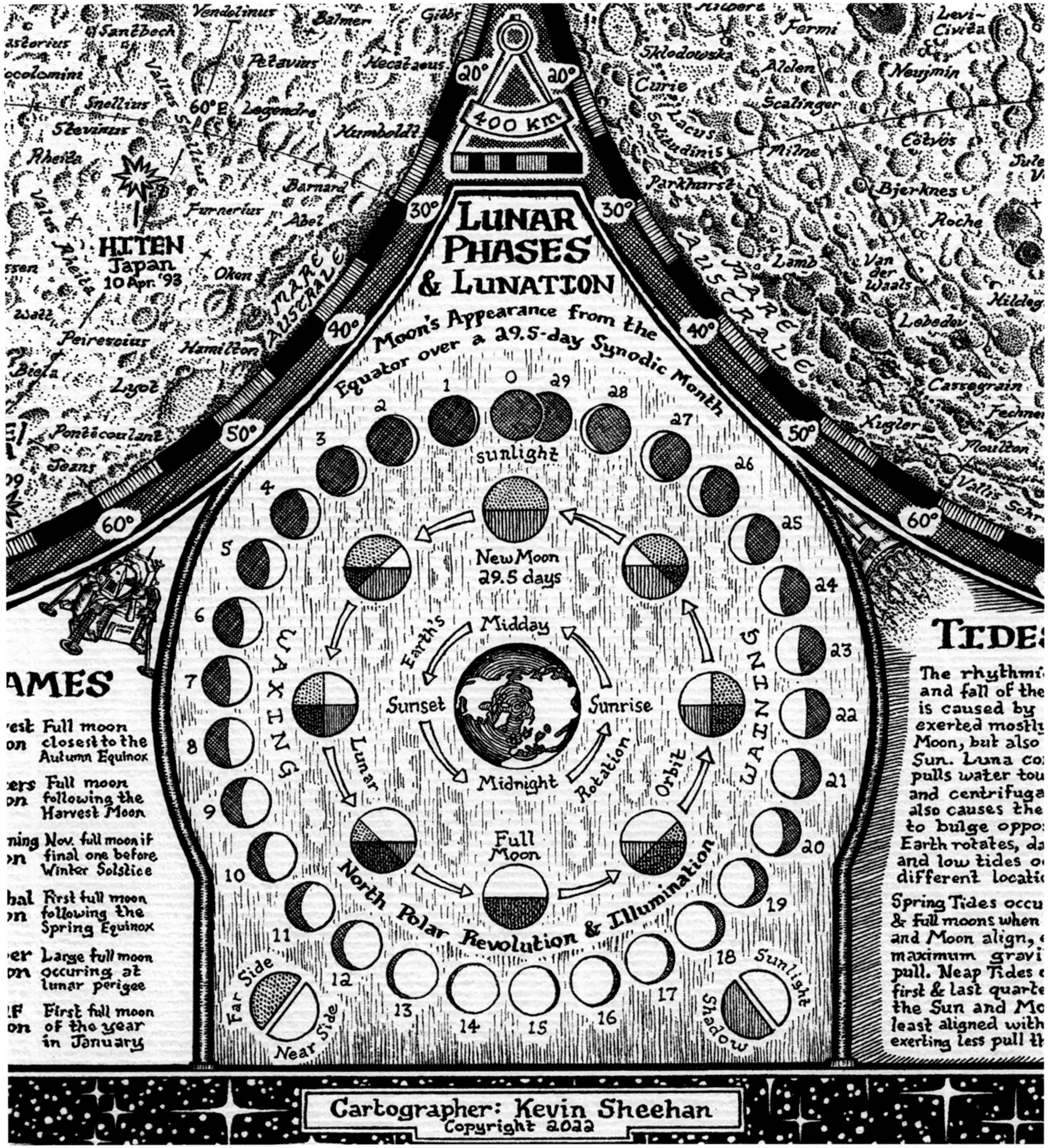
LUNAR PHASES & LUNATION
Moon's Appearance from the Equator over a 29.5-day Synodic Month
400 km
20°
30°
40°
50°
60°
sunlight
New Moon 29.5 days
Full Moon
Midday
Sunset
Sunrise
Midnight
Earth's Rotation
Lunar Orbit
WAXING
WANING
North Polar Revolution & Illumination
Far Side
Near Side
Sunlight
Shadow
0 1 2 3 4 5 6 7 8 9 10 11 12 13 14 15 16 17 18 19 20 21 22 23 24 25 26 27 28 29
Vendelinus
Balmer
Gibbs
Santbech
Petavius
Hecataeus
Snellius
Stevinus
Vallis Snellius
Legendre
Humboldt
Rheita
Furnerius
Barnard
Abel
Vallis Rheita
HITEN Japan 10 Apr. '93
Oken
MARE AUSTRALE
Hamilton
Peirescius
Biela
Lyot
Pontécoulant
Jeans
Fermi
Levi-Civita
Sklodowska
Curie
Alden
Neujmin
Lacus Solitudinis
Scaliger
Milne
Eötvös
Parkhurst
Bjerknes
Roche
Lamb
Van der Waals
Lebedev
Cassegrain
Kugler
Moulton
Vallis Schrödinger
AMES
Full moon closest to the Autumn Equinox
Full moon following the Harvest Moon
Nov. full moon if final one before Winter Solstice
First full moon following the Spring Equinox
Large full moon occuring at lunar perigee
First full moon of the year in January
TIDES
The rhythmic … and fall of the … is caused by … exerted mostly … Moon, but also … Sun. Luna co… pulls water tow… and centrifuga… also causes the … to bulge oppo… Earth rotates, da… and low tides o… different locati…
Spring Tides occu… & full moons when … and Moon align, … maximum gravi… pull. Neap Tides … first & last quarte… the Sun and Mo… least aligned with … exerting less pull th…
Cartographer: Kevin Sheehan
Copyright 2022

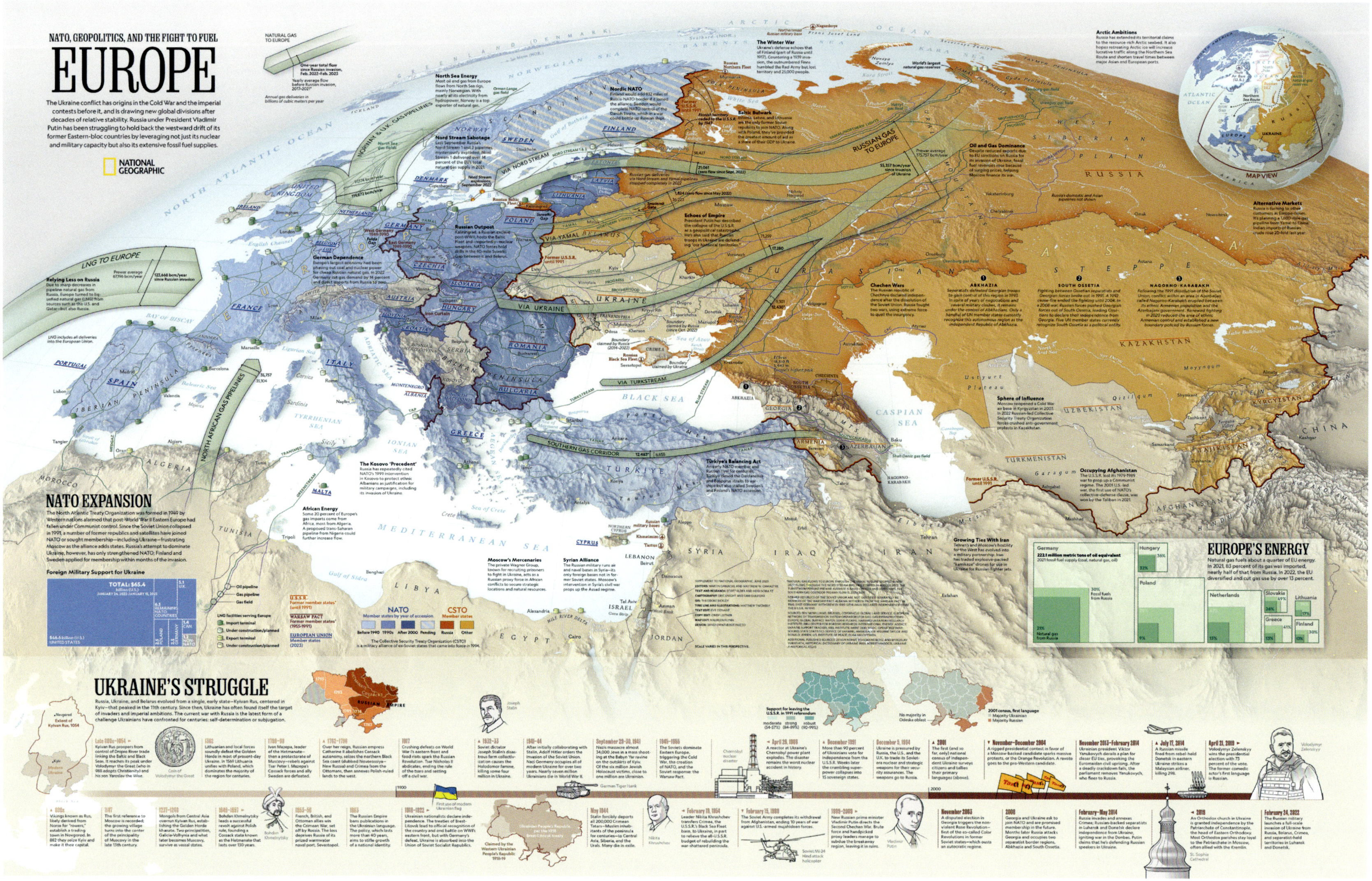
NATO, GEOPOLITICS, AND THE FIGHT TO FUEL
EUROPE
The Ukraine conflict has origins in the Cold War and the imperial contests before it, and is drawing new global divisions after decades of relative stability. Russia under President Vladimir Putin has been struggling to hold back the westward drift of its former Eastern-bloc countries by leveraging not just its nuclear and military capacity but also its extensive fossil fuel supplies.
NATIONAL GEOGRAPHIC
NATO EXPANSION
The North Atlantic Treaty Organization was formed in 1949 by Western nations alarmed that post-World War II Eastern Europe had fallen under Communist control. Since the Soviet Union collapsed in 1991, a number of former republics and satellites have joined NATO or sought membership—including Ukraine—frustrating Moscow as the alliance adds states. Russia's attempt to dominate Ukraine, however, has only strengthened NATO; Finland and Sweden applied for membership within months of the invasion.
Foreign Military Support for Ukraine
TOTAL: $65.4
$46.6 billion (U.S.) UNITED STATES
6.1 REMAINING NATO COUNTRIES
5.1 U.K.
2.5 POLAND
2.5 GERMANY
1.4 CAN
1.2 NON-NATO
Oil pipeline
Gas pipeline
Gas field
LNG facilities serving Europe
Import terminal
Under construction/planned
Export terminal
Under construction/planned
EUROPEAN UNION
Member states (2023)
NATO
CSTO
EUROPE'S ENERGY
Natural gas fuels about a quarter of EU energy. In 2021, 83 percent of its gas was imported, nearly half of that from Russia. In 2022, the EU diversified and cut gas use by over 13 percent.
Germany
Hungary
Poland
Netherlands
Slovakia 49% 26%
Lithuania 17%
Greece 41% 13%
Finland 30% 13%
UKRAINE'S STRUGGLE
Russia, Ukraine, and Belarus evolved from a single, early state—Kyivan Rus, centered in Kyiv—that peaked in the 11th century. Since then, Ukraine has often found itself the target of invaders and imperial ambitions. The current war with Russia is the latest form of a challenge Ukrainians have confronted for centuries: self-determination or subjugation.
Kyivan Rus prospers from control of Dnipro River trade linking the Baltic and Black Seas. It reaches its peak under Volodymyr the Great (who in 988 adopts Christianity) and his son Yaroslav the Wise.
Coin of Volodymyr the Great
Lithuanian and local forces soundly defeat the Golden Horde in most of present-day Ukraine. In 1569 Lithuania unifies with Poland, which dominates the majority of the region for centuries.
1708–09
Ivan Mazepa, leader of the Hetmanate—then a protectorate of Muscovy—rebels against Tsar Peter I. Mazepa's Cossack forces and ally Sweden are defeated.
ARCTIC OCEAN
NORTH ATLANTIC OCEAN
MEDITERRANEAN SEA
BLACK SEA
CASPIAN SEA
RUSSIA
UKRAINE
KAZAKHSTAN
VIA UKRAINE
VIA NORD STREAM
RUSSIAN GAS TO EUROPE
LNG TO EUROPE
SOUTHERN GAS CORRIDOR
NORTH AFRICAN GAS PIPELINES

« NATO, Geopolitics, and the Fight to Fuel Europe

Martin Gamache, Matthew W. Chwastyk, Scott Elder, Heidi Schultz, Eric Knight, Sam Guilford, Theodore Sickley, Matthew Twombly, Eve Conant, Cindy Leitner, Maureen Flynn, and Sandi Owatverot-Nuzzo, The National Geographic Society and National Geographic

This thematic map puts the Russia–Ukraine War into a larger geopolitical and historical context. As part of a foldout map supplement to National Geographic magazine, it complements a more conventional planimetric reference map of Ukraine, printed on the other side.

The overall design aims to harness the character and narrative power of hand-drawn, twentieth-century conflict maps, while taking full advantage of the capabilities of modern digital cartography and data sources. To stoke visual interest and help ground the subject matter in three-dimensional reality, a Richard Edes Harrison–inspired oblique view was chosen.

Once the map's scope was defined, we carefully composed a perspective view that captures the most relevant geography: European NATO and EU members, Russia and its allies, along with the former Soviet Union. The color scheme highlights these opposing alliances, reversing the conventional older-darker arrangement for NATO to build up contrast around Ukraine. The resulting negative space helps reinforce Ukraine's precarious position between Russia and the West.

The volumes of natural gas flow to Europe, before the war and since the war began, are shown as sweeping, semi-transparent arrows on a separate plane. These flow arrows are proportionally accurate, tapered to mimic the underlying land surface, and elegantly routed to minimize interference with other map content. To support these abstracted energy flows with depictions of actual infrastructure, we fused several incomplete pipeline datasets into a consistently generalized and stylized network.

The underlying geography is rendered with similar attention to detail. Natural landforms have been subtly and subjectively adjusted in order to actively reinforce and contextualize the primary content. Major rivers, for example, are carefully and legibly aligned alongside rather than beneath the array of overlapping political boundary lines. Terrain, coastlines, and the underlying sphere itself are tinted and shaded to maximize contrast and drama at the center of the map while deemphasizing the periphery.

Design details like these create a more legible map with less noise and less friction, helping to make this complicated subject matter accessible and attractive to a general audience without sacrificing seriousness or accuracy.

The Ukraine conflict has origins in the Cold War and the imperial contests before it, and is drawing new global divisions after decades of relative stability. Russia under President Vladimir Putin has been struggling to hold back the westward drift of its former Eastern-bloc countries by leveraging not just its nuclear and military capacity but also its extensive fossil fuel supplies.

NATIONAL GEOGRAPHIC

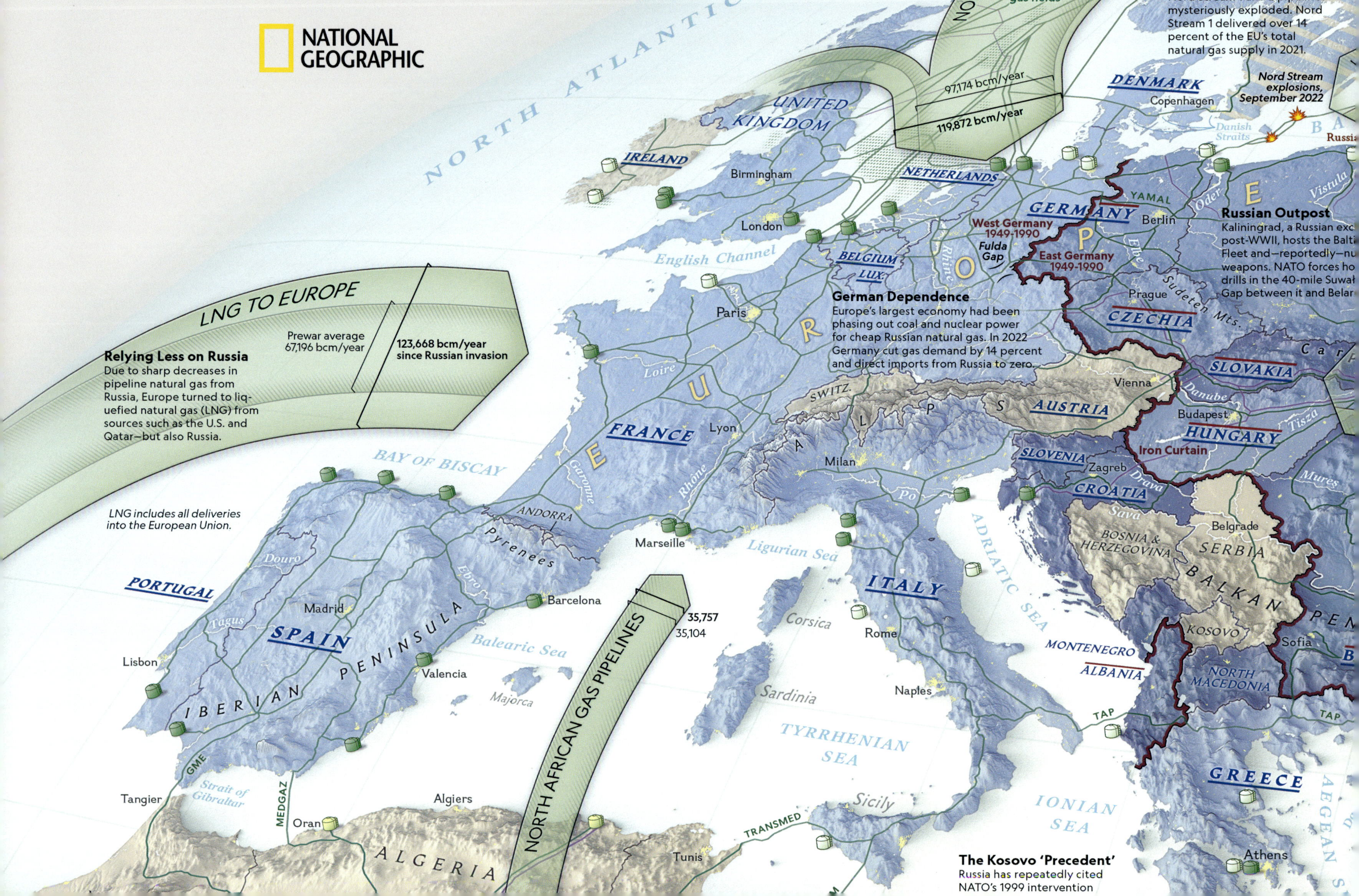

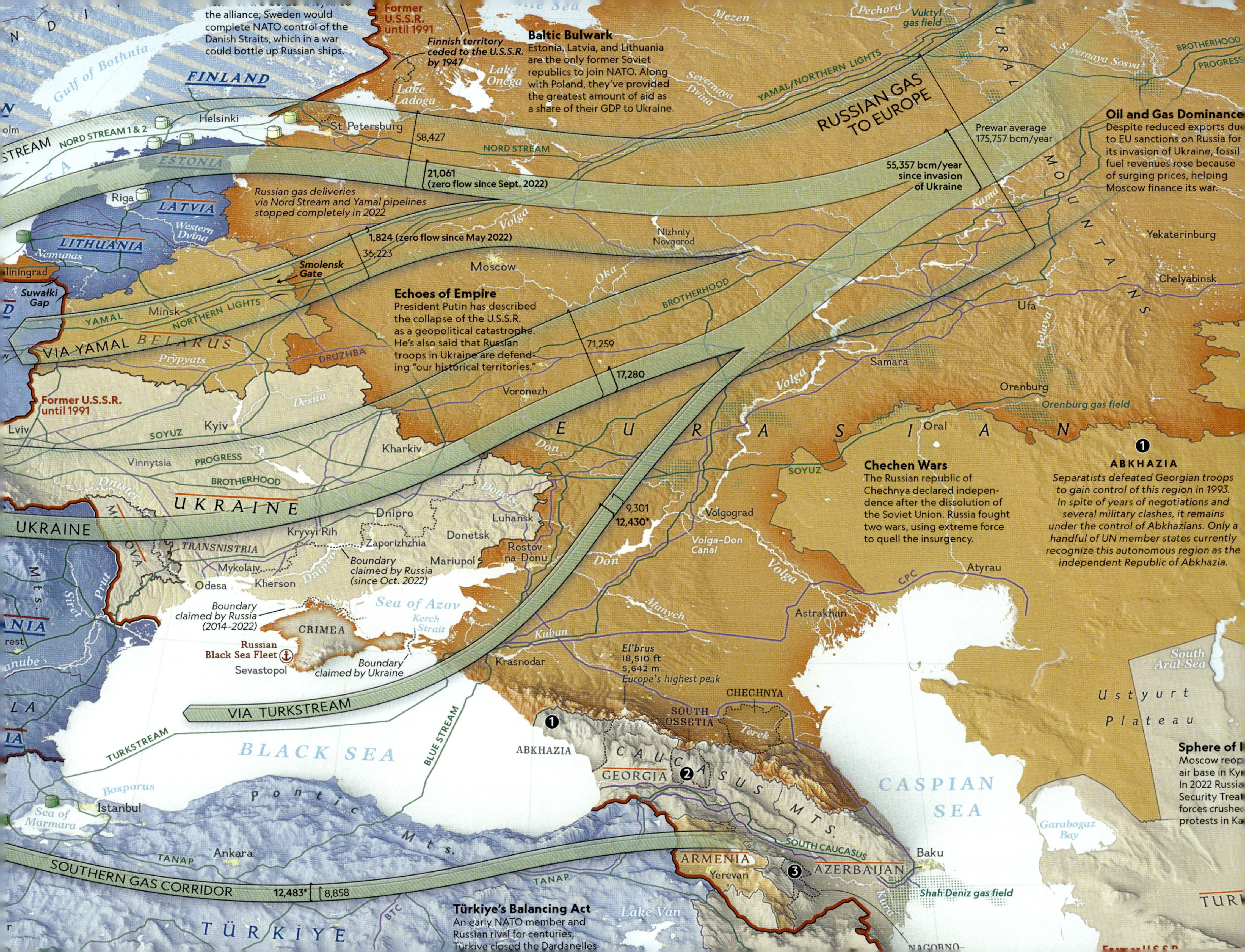

the alliance; Sweden would complete NATO control of the Danish Straits, which in a war could bottle up Russian ships.
Former U.S.S.R. until 1991
Finnish territory ceded to the U.S.S.R. by 1947
Baltic Bulwark
Estonia, Latvia, and Lithuania are the only former Soviet republics to join NATO. Along with Poland, they've provided the greatest amount of aid as a share of their GDP to Ukraine.
Gulf of Bothnia
FINLAND
Helsinki
St. Petersburg
Lake Ladoga
Lake Onega
Mezen
Pechora
Vuktyl gas field
Severnaya Dvina
URAL MOUNTAINS
BROTHERHOOD
PROGRESS
Severnaya Sosva
YAMAL/NORTHERN LIGHTS
RUSSIAN GAS TO EUROPE
Prewar average 175,757 bcm/year
55,357 bcm/year since invasion of Ukraine
Oil and Gas Dominance
Despite reduced exports due to EU sanctions on Russia for its invasion of Ukraine, fossil fuel revenues rose because of surging prices, helping Moscow finance its war.
NORD STREAM 1 & 2
NORD STREAM
58,427
21,061 (zero flow since Sept. 2022)
ESTONIA
Riga
LATVIA
Western Dvina
LITHUANIA
Nemunas
Kaliningrad
Suwałki Gap
Russian gas deliveries via Nord Stream and Yamal pipelines stopped completely in 2022
1,824 (zero flow since May 2022)
36,223
Smolensk Gate
Moscow
Nizhniy Novgorod
Volga
Kama
Oka
Yekaterinburg
Chelyabinsk
Ufa
Belaya
Minsk
YAMAL
NORTHERN LIGHTS
VIA YAMAL
BELARUS
Prypyats
DRUZHBA
Echoes of Empire
President Putin has described the collapse of the U.S.S.R. as a geopolitical catastrophe. He's also said that Russian troops in Ukraine are defending "our historical territories."
BROTHERHOOD
71,259
17,280
Samara
Orenburg
Orenburg gas field
Voronezh
Former U.S.S.R. until 1991
Desna
Lviv
Kyiv
SOYUZ
Kharkiv
Vinnytsia
PROGRESS
BROTHERHOOD
EURASIAN
Oral
Don
SOYUZ
Chechen Wars
The Russian republic of Chechnya declared independence after the dissolution of the Soviet Union. Russia fought two wars, using extreme force to quell the insurgency.
1 ABKHAZIA
Separatists defeated Georgian troops to gain control of this region in 1993. In spite of years of negotiations and several military clashes, it remains under the control of Abkhazians. Only a handful of UN member states currently recognize this autonomous region as the independent Republic of Abkhazia.
Dniester
MOLDOVA
UKRAINE
UKRAINE
Donets
Dnipro
Luhansk
Kryvyi Rih
Donetsk
Zaporizhzhia
9,301
12,430*
Volgograd
Volga-Don Canal
TRANSNISTRIA
Rostov-na-Donu
Mariupol
Boundary claimed by Russia (since Oct. 2022)
Mykolaiv
Kherson
Odesa
Dnipro
Don
Volga
Atyrau
CPC
Prut
Siret
Boundary claimed by Russia (2014–2022)
Sea of Azov
Kerch Strait
CRIMEA
Manych
Astrakhan
Russian Black Sea Fleet
Sevastopol
Boundary claimed by Ukraine
Kuban
Krasnodar
El'brus 18,510 ft 5,642 m Europe's highest peak
South Aral Sea
Ustyurt Plateau
VIA TURKSTREAM
CHECHNYA
SOUTH OSSETIA
Terek
TURKSTREAM
BLACK SEA
BLUE STREAM
ABKHAZIA
CAUCASUS MTS.
GEORGIA
CASPIAN SEA
Sphere of I
Moscow reop air base in Ky In 2022 Russia Security Treat forces crushe protests in Ka
Bosporus
Sea of Marmara
Istanbul
Pontic Mts.
Garabogaz Bay
Ankara
TANAP
SOUTH CAUCASUS
Baku
ARMENIA
Yerevan
AZERBAIJAN
SOUTHERN GAS CORRIDOR
12,483*
8,858
TANAP
Shah Deniz gas field
Kura
BTC
Türkiye's Balancing Act
An early NATO member and Russian rival for centuries, Türkiye closed the Dardanelles
Lake Van
TÜRKIYE
NAGORNO-

You've reached the end, but there's more to do,
One detail per map, each one with a clue,
The order is scrambled, the challenge is steep,
Revisit the maps for these things you must seek:

Four drainage basins fall short of the sea,
A little north arrow stuck up in a tree,

A painter's palette, and glacial Queen Mary,
A valley in the Keystone State called Cherry,

Twenty-four diamonds, BOOM! Ranger Nine,
A giant donut, blue tracks in a line,

Terre d'Ellsworth, a rose pointing East,
An Essential depiction of Illinois Peak,

A pair of Greats set side by side,
A delectable tart before you ride,

Out in the Wild: two red bulbous necks,
The tiniest island inside the complex,

Calgary's Palace, a lone power station,
In purple skies: Orion's constellation,

A jar in red, a pitcher in green,
An equine at work, two blasts in the Stream,

A flying kanuk, and plate twenty-five,
One of A.W. Perry's nine lives,

Palm trees in the foreground, Canterbury Bight,
A Cinema Arta no longer in sight,

Four mythical gates found off the coast,
A place in Tomatin for a warm toast,

At last Prater Park, and a beautiful bookstore,
We hope you enjoyed this excuse to explore more!

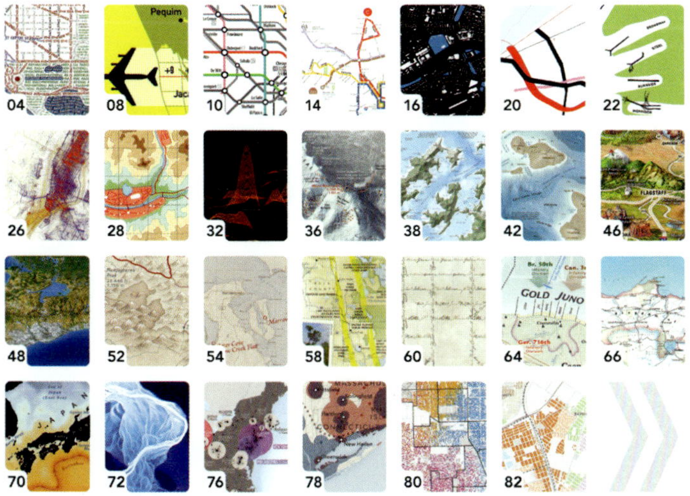

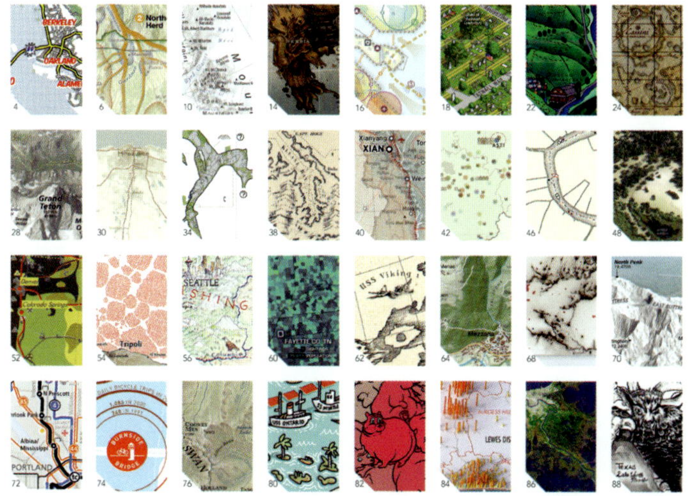

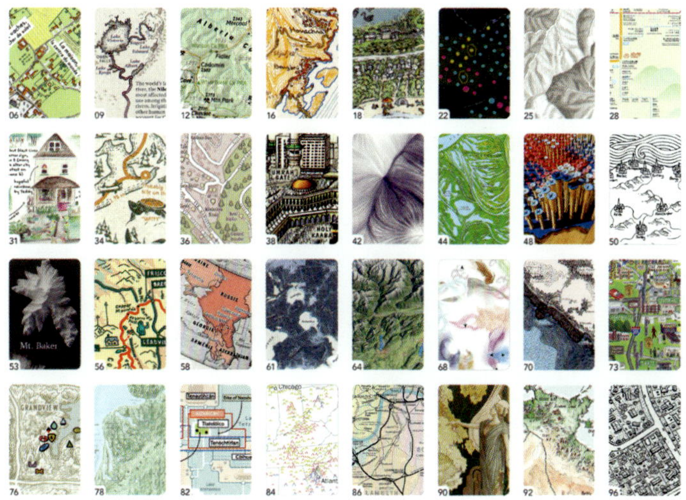

Volumes 1–7 available now at
atlasofdesign.org/order